MW00511163

# Job Order Contracting

**Expediting Construction Project Delivery**

By Allen L. Henderson

**RSMeans**

# Job Order Contracting

## Expediting Construction Project Delivery

- Comparing JOC to Other Delivery Methods
- Implementing a JOC Program
- Evaluating Proposals & Awarding Contracts
- General Requirements & Estimating
- The Advantages of JOC Partnering

**RSMeans**

Allen L. Henderson

 **Reed Construction Data**

Copyright © 2005
Reed Construction Data, Inc.
Construction Publishers & Consultants
63 Smiths Lane
Kingston, MA 02364-0800
781-422-5000
www.rsmeans.com
**RSMeans** is a product line of Reed Construction Data

Managing Editor: Mary Greene. Editor: Andrea Sillah. Editorial Assistant: Jessica Deady. Production Manager: Michael Kokernak. Production Coordinator: Laurie Thom. Composition: Sheryl Rose. Proofreader: Laurie Thom. Book and cover design: Sheryl Rose.

Printed in the United States of America.

10  9  8  7  6  5  4  3  2  1

Library of Congress Cataloging in Publication Data

ISBN 0-87629-811-0

# Table of Contents

# Acknowledgments

This book could not have been accomplished without the support of Robert Gair, Senior Consultant for RSMeans. He was instrumental in providing the opportunity for me to interact and partner in this undertaking with the many professionals at RSMeans. His support is deeply appreciated.

A special acknowledgment is extended to Texas State University-San Marcos. For over a decade, this institution has provided continued support for the JOC method of project delivery, endorsed its practical application, supported it with training and education, and experienced its beneficial results. My involvement with this program has enabled me to gather and assemble much of the information presented herein. I'd like to specially thank these facilities administrators for their past and present support: Ed Fauver, Allen Goldapp, Pat Fogarty, PE, and Coyle Buhler, PE. I am also grateful to Dr. Gary Winek for his recommendation to include delivery method comparisons, and to Dr. Stephen Springer for his helpful suggestions during manuscript development.

Much of the material that threads through this book was a direct result of assimilating information and experiences shared through various presentations by industry professionals and networking sessions among conference attendees during previous annual Texas Job Order Contract Conferences held at Texas State University-San Marcos. Recognition is also due to the Association of Higher Education Facility Officers and the Center for Job Order Contracting Excellence; their educational efforts with regard to JOC helped add to the content of this book.

Particular credit is due to presenters Robert Ayers, PE, and Ken Jayne, PE (whose work was helpful in developing the "History of JOC" section of Chapter 2); Joe Martin, PE; Gregory Smith, PE; and Kathleen Acock—all pioneers in the implementation and practice of the JOC delivery method. I would also like to recognize JOC presenters Robert Gair (cost estimating); Peter Giglio, AIA (delivery method overviews); Bill Tomlinson, RA, and Stephen Marlow (evaluating JOC RFPs); Angie Brown (JOC HUB opportunities); Bill Fly, Texas State Attorney, and Fernando Gomez, Vice Chancellor and General Counsel of Texas State University Systems (Construction Law); and many other presenters, owners, designers, contractors, and conference attendees too numerous to mention who have taken time to share their knowledge and experiences about JOC and interrelated disciplines.

Special recognition is due to Mary Greene, Manager of Reference Publications and Acquisitions Editor for RSMeans, especially with much-needed assistance in chapter definition and overall organization; Andrea Sillah, Manuscript Editor; and Jessica Deady, Editorial Assistant, for their invaluable assistance throughout manuscript development.

Many thanks also to Phil R. Waier, PE, Principal Engineer/Editor, and Robert W. Mewis, CCC, Senior Engineer/Editor, for their guidance in proposed book content at the onset of this venture, as well as their dedicated management, technical, and reviewing efforts. RSMeans' production and design staff offered significant contributions to the "Unit-Price Method" and "Resources" sections, as well as the "Estimating JOC" and "Anatomy of a Job Order" chapters.

Much appreciation is extended to the technical reviewers of this book, Ken Jayne, President of Applied Innovative Management, Inc. and a nationally recognized expert in job order contracting, and Rory Woolsey, construction consultant, estimator, and seminar presenter on JOC systems application. Their comments were very useful, adding key points to the book's content. Thanks also to Gary Aller for his contribution of the Foreword to this book. It presents an accurate account of current application and importance of JOC as a useful delivery method in today's construction industry.

The Cooperative Purchasing Network (TCPN) was also very helpful in providing information used in the coefficient development section.

And finally, a very special "thank you" to my wife and partner, Brenda Henderson, who tirelessly worked with me during the design and construction phase of this project, offering unwavering support of the effort, while developing most of the figures in this book.

# About the Author & Contributors

**Allen L. Henderson** is assistant director of facilities planning, design, and construction at Texas State University-San Marcos. He has over 30 years of experience in the construction industry as both a contractor and an owner's representative. During the past 25 years, Mr. Henderson has been responsible for many aspects of facilities services, including managing construction contracts for special projects at Texas State.

He has published articles, given numerous presentations, and consulted with various institutions on JOC methodology. He is the founder and president of the Texas Job Order Contract Conference, held annually at the university, and has over ten years of JOC contract management experience. He currently represents Texas State as a member of the Center for Job Order Contracting Excellence at Arizona State University.*

Mr. Henderson has also guest-lectured in cost estimating, model building codes, and ADA requirements and has been an instructor of industrial safety at Texas State. He holds accessibility and environmental credentials.

**Gary L. Aller,** who authored the foreword, is president and CEO of Strategic Business Concepts, LLC, a firm engaged in strategic planning, advising, recruiting, and consulting in the areas of business, operations, and program and project management. For the past 13 years, he has also worked for, and is currently the director of, the Alliance for Construction Excellence (ACE), which is part of the Del E. Webb School of Construction at Arizona State University. ACE serves as the national headquarters for the Center for Job Order Contracting Excellence (CJE).*

Mr. Aller's 33 years in the construction industry include project management and other senior management positions for the Bechtel Group of Companies, Ebasco, and J.B. Rodgers – Mechanical Contractors. His experience includes a broad range of work with different project delivery methods. His expertise is in business and operations management, program and project management, research, recruiting, engineering and construction management, retrofit and maintenance management, contract/subcontract management, labor relations, cost/schedule controls, craft supervision, and start-up and commissioning.

**Ken Jayne** is the president and CEO of Applied Innovative Management (AIM), a management consulting firm that helps facility owners select and manage best-value service providers. A nationally recognized expert in the field of job order contracting, Mr. Jayne was a founding member, first industry chair, and chairman of the Education Committee of the Center for Job Order Contracting Excellence.* He developed the TCPN AJOC<sup>sm</sup> program to enable small school districts to obtain responsive, cost-competitive, high-quality JOC construction services.

Mr. Jayne has more than 40 years of experience in facilities management, business and organizational development, operations, and engineering and served 27 years in the U.S. Army Corps of Engineers. His last military position was director of public works at Fort Hood, Texas. Prior to his position with AIM, Mr. Jayne managed the JOC BD team for a large engineering and construction company.

**Phillip R. Waier** is a principal engineer at RSMeans and senior editor of the annually updated publications *Facilities Maintenance & Repair Cost Data* and *Building Construction Cost Data*. He manages the activities of editors, cost researchers, and consultants and delivers consulting services to public and private agencies and firms throughout the United States.

Mr. Waier's 30 years in the construction industry include positions as president and chief engineer of a mechanical contracting firm, project manager for numerous industrial construction projects, and structural engineer for major foreign and domestic construction projects. He is a registered professional engineer and a member of the Association for Facilities Engineering (AFE), Associated Builders and Contractors (ABC), and Associated General Contractors (AGC).

**Rory M. Woolsey** has over 30 years of experience in construction management and engineering with firms including Bechtel, H.J. Kaiser, and RSMeans. He has held positions as field engineer, project manager, MIS manager, testing laboratory manager, estimator, senior editor, designer, structural engineer, and general contractor. He has consulted in the job order contracting environment for 18 years, building coefficient models, researching factors for area-specific coefficients, estimating for delivery orders, writing proposals, and as a site manager of a JOC contract averaging $6 million per year in delivery order work volume.

Mr. Woolsey has presented over 6,000 classroom hours of instruction on construction estimating, JOC systems application, partnering, project management, and the application of the RSMeans database to many organizations (including site-specific training for the U.S. Navy, Air Force, and the Air National Guard). He is a member of the Association for the Advancement of Cost Engineering.

*CJE is a combined industry, facility-owner, and academic organization under the Alliance for Construction Excellence, Arizona State University, dedicated to improving the JOC industry, educating customers and clients, and developing more effective performance-based methods for contractor selection.*

# Foreword

*Gary L. Aller, Director*
*Alliance for Construction Excellence*
*Del E. Webb School of Construction*
*Arizona State University*

Have you ever felt trapped by the small project dilemma? In this situation, there are too many small things to be done—repairs, maintenance tasks, renovations, or small construction projects. Yet none are big enough individually to attract anyone's attention or to justify the administrative time and expense required for the standard project procurement procedures. These are the projects that never seem to leave the bottom of your task list, taking a low priority beneath the bigger jobs. The fact remains that these small jobs are no less important to those who need to have them accomplished.

Fortunately, there is a cost-effective, time-efficient solution. It is called job order contracting, or JOC. This book will give you all the information needed to use this flexible project delivery method to its fullest potential. Job order contracting relies on pre-established unit prices. It provides an owner with an on-call contractor who is familiar with the site and the owner's needs. JOC is a vitally important tool in the owner's contracting repertoire.

Standard medium-sized and large construction projects that result in new facilities can be built efficiently using delivery methods such as design-build or construction manager-at-risk. With these larger projects, the process of planning, soliciting qualifications and proposals, and conducting competitive evaluation of firms does not constitute a large percentage of total project cost or time expended. For small projects, however, the procurement process consumes a larger chunk of the project budget if the projects have to be contracted individually every time. The result is increased total project cost and completion time.

The JOC delivery method eliminates this problem. It is particularly well-suited to repetitive jobs and small tasks that will need to be accomplished over a period of time, without the exact timing and detailed requirements of each job completely defined. Many diverse tasks, such as routine maintenance, upgrades and renovations, alterations, and minor construction, are handled efficiently using a single JOC contract.

Independent research studies have shown significant cost savings associated with the use of JOC. These savings are a result of lower procurement costs per project, lower contractor overhead per project, reduced design costs, and fewer change orders and claims.

Despite the documented benefits, misconceptions and a misunderstanding of the JOC process have prevented this valuable method from being used as often as it could be. The most common myth is that the process involves no competition, resulting in higher costs. A frequent misconception is that once the contract is in place, there is no control over pricing and the contractor can inflate costs. Another is that JOC projects are not open for scrutiny.

The reality is that job order contracting has been used with great success for more than a decade in both public and private sectors. The results have shown that JOC contracts provide better quality work with lower project costs and shorter completion times. This book clearly dispels the myths surrounding job order contracting and contributes to a better understanding of the process.

Once accurate information about the nature of job order contracting is understood, it is easy to see the important advantages it provides. First and foremost, each job is not subject to the delays and costs associated with the traditional procurement process. Equally important is the long-term relationship with a quality job order contracting firm, which encourages efficient communication, effective teamwork, and mutual trust. These established partnerships result in faster service and higher quality and are the key to realizing the most benefit from the job order contracting experience.

The importance of addressing smaller projects and their role in overall facility function should not be underestimated. Many of the projects that are efficiently and cost-effectively accomplished using job order contracting are vital to the future use of existing facilities. Maintenance delays lead to safety problems and shortened facility life, and deferring renovations and improvements detracts from the facility's usefulness. Small projects are the key to getting the most for your money in terms of facility usability and longevity. This philosophy places job order contracting in an important position in the industry and emphasizes the essential and timely contribution made by this book.

# Preface

The conference presentation* was intriguing and a little unorthodox. The topic was a relatively new construction project delivery method called delivery order contracting (DOC), which was successfully being administered at Texas A&M University-College Station. The method was introduced as an innovative contracting tool for facility owners, effective for minor construction, renovation, and rehabilitation projects.

According to the speaker, complete bid packages with detailed design documents were no longer needed for each project. Construction costs could be projected accurately, and each project would no longer require a lengthy and costly competitive bidding process. A new project would not necessarily require a new contractor to become familiar with our institution's contract provisions, our policies and procedures, our facilities—*our needs and expectations.* It all sounded too good to be true…

Over a decade has passed since that conference. As a higher education public facility planner/estimator, I hardly realized then the ramifications that were to follow from that event. It marked the beginning of a journey towards research, implementation, and practice of this unique project delivery method, now more commonly referred to as job order contracting (JOC). The benefits, pitfalls, and intricacies would be shared among others who were also gaining experience through active participation in this contracting method.

The primary objective of this book is to introduce JOC to current and future construction and facilities professionals who seek fundamental information about its distinctive contract provisions and procedures.

This book is also intended for those professionals who are contemplating JOC participation and need specific guidelines for successful program implementation and execution.

JOC is more than just another construction project delivery method. A successful JOC program consists of long-term, collaborative processes that provide a unique opportunity among program participants to develop a mindset of partnering—a strong trend in the construction industry that begins with initial project design and continues throughout the construction process. Partnering has been proven to reduce the likelihood of adversarial relationships, thereby reducing risks and promoting the achievement of mutually beneficial goals.

Construction contract provisions and administration procedures have evolved over time, leading to the current trend toward involving the general contractor or construction manager early in the design process. Innovative contract provisions promote procedures that enable accelerated project start dates, as well as a reduction in the duration of individual construction sequences.

Contractor award criteria are shifting away from just "low bid" to include a more thorough evaluation of a contractor's qualifications and capabilities, including a firm's ability to offer services such as technical assistance and cost estimating during the design phase. Award procedures include contractor pre-qualification followed by owner-conducted interviews. Although project pricing is always a consideration for both facility owners and contractors, JOC contracts are requirements-based, with performance as a primary factor, in alignment with current industry trends regarding contract award to contractors offering the owner the best overall value.

Other recent trends include strategies that address risks inherent in the industry—especially the practice of cost-capping projects as early as possible during the design phase to curb ever-escalating material prices, labor rates, and service fees. Mandates for increased security, building life-cycle energy usage, and environmental impact analysis will continue to guide future design efforts. Ongoing improvements in tool technology, new composites, "green" materials and building systems, and computer software applications for construction cost estimating and scheduling will continue to impact the industry. JOC is at the forefront of many of these trends.

The JOC contract identifies compensation to the contractor based on competitively bid multipliers, or *coefficients,* applied to pre-established unit-pricing provided by the industry's most reputable sources—using state-of-the-art cost estimating databases capable of linking to product manufacturers' Web sites. This process enables proficiency in material and product selection and allows a high degree of accuracy in estimating individual project costs—all carried out in an expeditious manner. The result is that JOC participants are better able to execute large volumes of work quickly and within established budgets.

Construction today is truly a dynamic, exciting industry. Current trends align perfectly with the concept and principles of JOC, as its effectiveness continues to improve through proper implementation and practice. There is no question that the method is gaining respect among industry professionals. It is emerging and steadily taking its place among other recognized construction project delivery methods as the best choice among both facility owners and contractors for the types of projects suited to its application. The seed has been firmly planted.

The publisher and the author have jointly recognized a need to develop a comprehensive publication that takes into account currently recognized aspects of JOC, explains how JOC best fits into the construction industry, and includes instructional content regarding method implementation and contract execution procedures. My efforts in this endeavor have been to summarize and present a compilation of informative contributions from industry professionals who have been willing to expend their valuable time, sharing their knowledge, experiences, insights, and passion for JOC with others.

*The conference was held by the Association of Higher Education Facility Officers (APPA) in San Antonio, Texas.*

# Chapter 1

# A Review of Project Delivery Methods

*"The most damaging phrase in the language is 'It's always been done that way.'"* — Grace Murray Hopper, Ph.D., computer pioneer and Rear Admiral, United States Navy

Each of the two phases of a project—design and construction—consists of a sequential series of interrelated processes that are influenced by time, cost, and quality. The choice of project delivery method can directly affect the overall time line and cost of a project—and has the potential to influence the working relationships among project participants, thereby affecting the quality of their performance.

An owner chooses the project delivery method that is most advantageous to a particular project. The selected method is a contracting "tool" that will be used to administer the project's construction phase and, with some methods, the design phase.

Until recently, most project delivery methods fostered only process-oriented and, in a sense, distant relationships among project participants. These traditional methods involve selection and award of professional design services (to develop comprehensive, complete design documents), followed by a separate process for construction services to accomplish and deliver the project to the owner. These are commonly referred to as *design-bid-build* (DBB) type methods and are still in use today.

Currently, owners have several options other than traditional DBB methods. These delivery method alternatives promote interaction among the owner, the design phase participants, and the construction phase participants. These approaches have gained popularity in the construction industry, primarily because they can accelerate pre-construction time lines, but they offer other attributes as well. Job order contracting, or JOC (also referred to as delivery order contracting or DOC), is one such method.

JOC is well-matched to meet many of the project delivery needs of today's facility owners—particularly owners involved in public education; municipalities; local, state, or federal agencies; and the military, as well as other entities with facility project needs, both public and private.

This chapter will review alternate project delivery methods currently in practice and accepted by the construction industry:

- Job Order Contracting
- Competitive Sealed Bid & Proposal
- Cost Plus Fee & Guaranteed Maximum Price
- Construction Manager & Construction Manager-at-Risk
- Design-Build
- Time & Materials
- Unit Price

Comparing these delivery methods and their features will enable owners to better select the right method for a given project. The discussion of JOC identifies its primary attributes. Subsequent chapters of this book will build on this groundwork and discuss the facets of JOC in depth and how to implement it.

## Job Order Contracting

Job order contracting is an ideal project delivery method for minor construction, renovation, rehabilitation, and maintenance projects. The quality of work performed is usually equal to or greater than any other project delivery method currently in use. Project costs are usually equal to or less than other methods, and the criteria used for pricing is typically firm, objective, and consistent.

JOC's pricing structure provides consistently accurate and predictable project cost estimating, and the method can be used for construction work as well as for facilities maintenance or specific trade groups. In addition to these attributes, JOC contractors are service-oriented to the owner.

JOC is a perfect match for owners who have the need to complete multiple small- to medium-sized repair and renovation projects easily and quickly. Once a JOC contract is in place, the owner simply identifies each project with a brief description and notes the desired or required dates and times for performing the work. The JOC contractor is notified by the owner, who requests a design (if necessary), a detailed scope of work, and a price proposal for the project. This process is commonly referred to in JOC as a *Request for a Job Order (JO) Proposal.*

The owner, JOC contractor, and designer (if necessary) work closely during the site visit to identify site characteristics and decide on the most economically advantageous means and methods needed to perform the work. The design (if needed), along with a detailed scope of work including the project's performance times, is then submitted by the JOC contractor to the owner for consideration. Once these submittals are mutually agreed on, the JOC contractor submits a detailed, lump-sum fixed price proposal based on the defined scope of work.

The combination of these submitted documents is referred to as the JOC contractor's *job order proposal*. The owner reviews the price proposal for accuracy in accordance with the JOC contract provisions. When the JO proposal approval process is completed, the owner signs off on the proposal and the project can proceed as scheduled.

## Indefinite Delivery/Indefinite Quantity

Owners often have a need to complete multiple projects, sometimes simultaneously. In these cases, the traditional DBB project delivery method can be too slow to handle the project volume with expediency. A JOC contract is ongoing and remains intact project after project. The overall quantity, or number of projects or tasks to be delivered, is indefinite. Projects that are yet to be determined by the owner can be administered through the use of the contract during the duration, or *term*, of the contract. These projects may be indefinite as to start and completion dates, as well as indefinite regarding the scope of work and associated unit-priced quantities. The inclusion of these characteristics in contract provisions identifies the JOC contract as having *indefinite delivery/indefinite quantity (ID/IQ)* features.

## Design Documents

If design documents are either required or desired, they do not have to undergo the extensive level of effort and painstaking detail that is required with the traditional, competitively bid design-bid-build delivery method. They need only enough detail for the contractor to understand the owner's intent, obtain subcontractor quotes and permitting, if required, and/or comply with applicable regulatory requirements. JOC practitioners refer to these drawings as *code review drawings,* or sometimes *incidental drawings,* and, in many cases, they can be developed by or through the contractor. They are incidental in nature—supplementary companions to a written scope of work. Often, design documents are not needed at all, especially if the project does not require record drawings to document regulatory compliance and the scope of work sufficiently documents the work to be performed.

The level of design effort, in conjunction with the procurement process for the project delivery method chosen, directly affects the pre-construction time line of the project.

## Comparing JOC with Traditional Methods

Figure 1.1 compares pre-construction time lines using JOC to the traditional DBB delivery method. The traditional method includes professional architectural/engineering (A/E) design services by award through a Request for Qualifications (RFQ) and negotiated fee. The JOC method includes code review drawings through the JOC contractor. It is not unusual for JOC pre-construction time lines to average 10–15 days if the scope of work is clear, the work is of a repetitive nature, and materials specified for the project are readily available.

With JOC, numerous projects can be in progress concurrently under one contract, allowing the owner to complete a high volume of projects as the need arises (subject to the contractor's capabilities). The method is a good candidate for projects that require phasing to accommodate

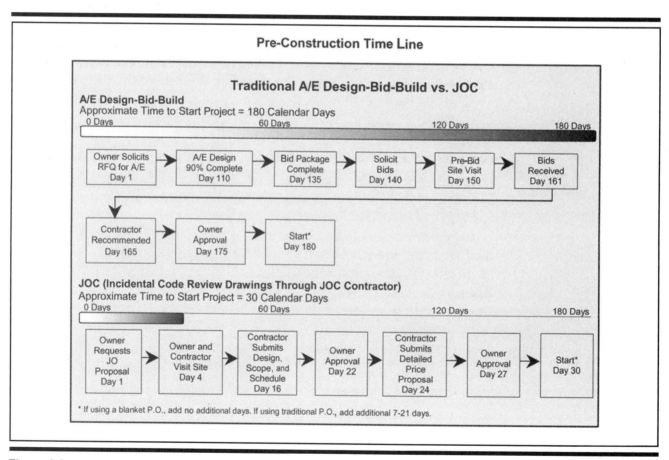

Figure 1.1

**Pre-Construction Time Line**

**Traditional A/E Design-Bid-Build vs. JOC**

**A/E Design-Bid-Build**
Approximate Time to Start Project = 180 Calendar Days

0 Days — 60 Days — 120 Days — 180 Days

- Owner Solicits RFQ for A/E — Day 1
- A/E Design 90% Complete — Day 110
- Bid Package Complete — Day 135
- Solicit Bids — Day 140
- Pre-Bid Site Visit — Day 150
- Bids Received — Day 161
- Contractor Recommended — Day 165
- Owner Approval — Day 175
- Start* — Day 180

**JOC (Incidental Code Review Drawings Through JOC Contractor)**
Approximate Time to Start Project = 30 Calendar Days

0 Days — 60 Days — 120 Days — 180 Days

- Owner Requests JO Proposal — Day 1
- Owner and Contractor Visit Site — Day 4
- Contractor Submits Design, Scope, and Schedule — Day 16
- Owner Approval — Day 22
- Contractor Submits Detailed Price Proposal — Day 24
- Owner Approval — Day 27
- Start* — Day 30

* If using a blanket P.O., add no additional days. If using traditional P.O., add additional 7-21 days.

4

operations and/or budget constraints, and is particularly well-matched for projects with critical performance times. Additionally, having a JOC contractor already mobilized at a facility or in the surrounding area can give owners the capability of almost immediate mobilization for emergency projects.

## Quality

**The quality of work performed with a JOC contract is usually equal to or greater than that of other project delivery systems.** When a contractor submits a proposal for a JOC contract, the proposal evaluation for award is thorough and based heavily on the respondent's past performance, qualifications, and project management capabilities. Once awarded, consistent high-quality work is essential to the longevity of the JOC contractor's relationship with the owner—and the overall success of the contract. The contractor becomes familiar with the owner's requirements and expectations and strives to meet them.

These factors help ensure that the quality of work performed will be equal to or higher than that of other project delivery methods. However, less-than-perfect quality in workmanship and/or materials is sometimes desired in order to match existing construction and meet budget constraints and/or project time lines. If so, the agreed-on scope of work can define when this approach is applicable to a specific project or task—exhibiting one of the flexibility features of JOC.

## Cost

**With JOC, project costs are usually equal to or less than those of other project delivery methods.** Traditional delivery methods for general contracting work are project-specific. Consequently, the contractor's estimated overhead costs for main office expenses, field office expenses, materials storage, project management, and full-time dedicated field supervision are included in the bid or price proposal for the specific project solicited for award. With JOC, these costs are spread out over numerous projects over a long period of time.

The JOC contractor's project management and supervisory field personnel, once familiar with the owner's requirements and expectations, do not need to be dedicated to only one project at a time—especially if the projects are small in scope and scale. (Of course, a contractor's representative should always be required to be present or readily available at each project site—but need not always be dedicated to supervision only with JOC projects.) With this distribution of the JOC contractor's labor burden, the more volume of work the JOC contractor receives from the owner, the fewer overhead expenses the contractor will have to absorb per project. The result is cost savings for the contractor that can be passed on to the owner.

With project-specific delivery methods, contractors are typically awarded only a small percentage of the contracts for which they submit bids or proposals, usually 15%–25%. The contractor's effort to submit a bid or proposal to the owner for consideration of contract award is extensive, and therefore expensive. This effort is lost if the contractor is not awarded the contract. Contractors usually average their costs for such lost efforts as increases to their main office overhead and add this amount to subsequent bids or price proposals.

With JOC, the contractor's price proposal offered and accepted in response to contract solicitation establishes a set criterion for pricing to be used for each project authorized during the term of the contract. The pricing for each project is submitted to the owner for approval in accordance with this pre-established criterion. Because each project does not undergo traditional competitive bidding, the result is cost savings for both the owner and the contractor.

The JOC arrangement typically yields a stable relationship between the contractor and owner, as well as between the contractor and subcontractors. Like contractors, subcontractors traditionally are awarded only a small percentage of the work for which they estimate and submit bids. They will usually give JOC contractors favorable pricing in exchange for a more reliable volume of work.

These factors have the potential for real cost savings, associated mostly with the contractor's reduction in overhead expenses. For projects using traditional delivery methods, contractors typically apply a percentage or set fee in relation to each project they bid. With JOC, the profit margin is preset within the pricing structure of each job order and is related to the quantity of each unit of work identified to complete the project. Therefore, a JOC contractor's profit margins are tied directly to the volume of work received, just as overhead is reduced by that volume. In a healthy, competitively bid atmosphere, the JOC contractor's proposed price to the owner should reflect these savings, relative to the annual volume of anticipated work.

Additional cost savings to the owner through JOC can be achieved by the following:

- Eliminating or reducing design efforts
- Eliminating project-specific bid or proposal documents and the expense of their reproduction and distribution
- Eliminating advertising costs for invitations to contractors to submit project-specific bids or proposals
- Increasing the potential for significant savings (as compared to the DBB method) via value engineering alternatives and

technical assistance from the contractor during the design and/or scoping phases

- Reducing contract administration efforts as the owner's relationship with the JOC contractor develops over time

## JOC Pricing Structure

**The pricing criterion for JOC is firm, objective, and consistent.** Once a project's scope of work is established and design documents are developed, the owner-approved price proposal becomes a *lump-sum fixed price* for the agreed-on scope of work. The fundamental provisions for compensation under a JOC contract are unit price-based, but result in a lump-sum, fixed-price proposal once the parties have agreed on the scope of work and the appropriate unit-price line items with associated quantities. JOC relies heavily on consistent, mutually accepted unit prices and estimating practices.

The components of a JOC price proposal are derived from reliable, impartially developed, pre-established unit cost data, combined (in what is commonly referred to as the *unit-price book, or UPB*) with software applications. This cost data comes from reputable sources along with competitively bid cost modifiers *(coefficients)*—resulting in fair compensation to the contractor. Since both the unit-price data and the coefficients are set at the time of contract award, the criteria for contractor compensation is consistent throughout the term of the contract, even if there are changes or modifications to an authorized job order.

The pricing structure for JOC is in direct contrast with the pricing structures of other delivery methods. This is because other methods use contractor-established pricing, based to a large extent on subcontractor pricing. Even if evaluated by the contractor through competitive bidding, subcontractor pricing has the potential to be unstable and inconsistent, reflecting wide variances in overhead, profit margins, and labor productivity rates. JOC contracts allow contractor compensation based directly on subcontractor pricing *only* when specifically authorized by the owner due to special circumstances.

## JOC as Applied to Maintenance

**Facility maintenance and other needs can be administered easily through a JOC program.** JOC projects and tasks are limited only by the availability of unit prices designated for use in accordance with the provisions of the contract, along with the capabilities of the JOC contractor. A well-chosen UPB or a designated group of UPBs will provide standard unit prices for a wide variety of tasks, and JOC contracts provide mechanisms to address items not found in designated UPBs. *(See "Non-Prepriced Work" in Chapter 5.)* This means that JOC can be utilized for almost any task associated with new construction, renovation, alteration, special projects, or maintenance, with approval of the administrative and purchasing staff who have jurisdiction.

Although JOC is typically earmarked for general construction applications, a contract can be developed to accommodate facility maintenance needs, groundskeeping, or landscaping, either as separate contracts or as *additive alternates* to a single contract. JOC contracts developed for specific trades or trade groups have the potential to offer favorable pricing to owners since they are contracting directly with installing contractors. However, if use of JOC for these non-construction needs is anticipated, the owner should indicate this, along with an estimate of the potential magnitude in the Request for Proposal (RFP), since the relationship of actual pricing to unit prices for these items may be significantly different than for construction-related items. A sufficient amount of owner staffing is required to provide administrative support for any JOC program. Such staffing must be taken into consideration when evaluating the possible implementation of auxiliary JOC programs.

## Service
**JOC contractors are often more service-oriented than contractors engaged in more traditional delivery methods.** Since their profit margins are related to the quantity of work received and the contract allows for an indefinite quantity of work, JOC contractors realize that future work will continue if good relations can be maintained with an owner who is financially stable and has a repetitive need for JOC services. In fact, most successful JOC contractors are highly motivated to make service to the owner (such as timely response to requests, accurate proposals, and quality workmanship) a top priority so that additional work will continue to flow their way.

## The Evolution of JOC
JOC is a flexible project delivery method with evolving attributes. Its implementation has expanded beyond its original methodology. Diverse practices are coming into play that enhance its benefits—substantial reductions in project bid-to-completion time, better value for the construction dollar, better contractor/client relationships, and high levels of end user satisfaction. JOC clearly passes the test of public-sector procurement requirements. It involves thorough evaluation of an RFP respondent's qualifications and performance abilities. This, together with competitive pricing, provides the owner with the best overall value.

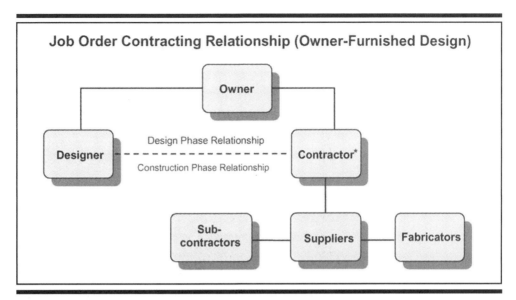

**Figure 1.2a** illustrates the JOC relationship when the owner furnishes the design. Owner/professional designer, owner/contractor, and contractor/subcontracted resources contractual relationships are represented by solid lines. The professional designer/contractor relationship is process-oriented, not contractual (dashed line). However, the owner, JOC contractor, and designer have a mutual incentive to work closely throughout the design and construction phases. The requirement for the designer and the contractor to assist each other may be contractual with the owner. Optional: design services may be performed with the owner's in-house designers.

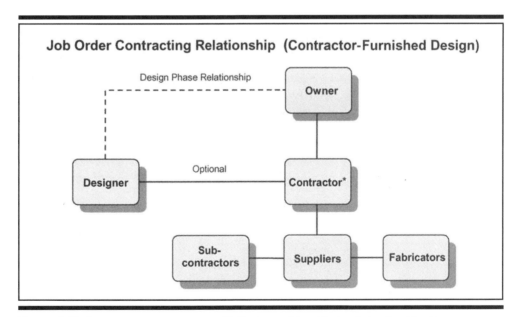

**Figure 1.2b** illustrates a more common JOC relationship with the design furnished by or through the JOC contractor. Owner/contractor, contractor/professional designer, and contractor/subcontracted resources contractual relationships are represented by solid lines. The design professional, if desired or required, is subcontracted by the JOC contractor as an option, working closely with the owner and the contractor through the design phase. In most cases, specific project design efforts are developed by the JOC contractor working closely with the owner and include sufficient content to reflect code compliance.

*\*Note: Although JOC primarily uses subcontractors rather than the contractor's in-house workforce, frequently the use of the contractor's in-house workforce is desired (or required).*

# Competitive Sealed Bids— Stipulated Sum

Also referred to as *lump-sum* or *fixed-price contracts,* these have traditionally been the standard choice among facility owners, and are commonly used for the delivery of most all types of projects. The contractor submits a "bid," identified as the total sum of money to be paid to the contractor for the entirety of work to be performed to accomplish the project. The owner then selects from a group of competitors a single contractor whose stipulated sum is most advantageous to the established project budget.

Stipulated sum delivery methods can be well-matched to large-scale projects of complex or innovative design when it is advantageous that the design of the project is completed in its entirety prior to the construction phase. The lump-sum feature lets the owner know the apparent cost of the project before the construction phase begins. However, since there is a probability of change orders and claims, a contingency allowance is normally programmed into the total project budget. (Very seldom is the "bid" price the final construction cost.)

With the stipulated sum method, the contractor serves as the single point of responsibility to the owner for the total construction process. This includes providing construction management services, such as scheduling, coordination, and oversight of the work; gathering bids from subcontractors; and engaging in contractual relationships as needed. All labor, materials, and equipment are traditionally provided by the contractor. The contractor is also responsible for executing all the various project tasks, monitoring security of the project site and safety of the workers, and rectifying any physical damages associated with the project.

The owner and contractor may be jointly responsible for public safety and environmental impact issues. Other responsibilities and obligations of both the contractor and the owner should be clearly outlined in the contract documents.

## Design & Bid Documents

### Design Documents

Before the bidding process can begin, the owner must select a design professional. If funding for the project has not yet been established, the initial design effort may be limited to conceptual renderings with associated cost estimates. This preliminary information enables the owner to ensure financial support for the project. When funding is secured, the design phase can continue.

Once produced, the design documents are coupled with General and Supplementary Conditions and other related documents, which together comprise the project's *bid documents.* General Conditions (sometimes referred to as "boilerplate" documents) delineate the basic provisions of most construction contractual agreements.

Supplementary Conditions complement the General Conditions and modify their provisions through specific changes, deletions, or additions as desired by the owner. The compiled bid documents are then advertised in accordance with public-sector procurement requirements, if applicable. This solicitation, or posting, typically identifies a date for interested contractors to attend a *pre-bid conference.* Sometimes owners may require mandatory attendance at pre-bid conferences, where prospective contractors can become familiar with the project's site conditions. These meetings also provide a forum for questions and answers with the designer and/or owner, as well as other project participants.

**Comparison of Features: Stipulated Sum vs. JOC**

| Delivery Method Features | Delivery Method | |
|---|---|---|
| | **Stipulated Sum** | **JOC** |
| **Contract type:** | Project-specific | ID/IQ |
| **Term of contract:** | Project-specific | 1 year + options (typical) |
| **Primary application:** | Multi-trade | Multi-trade |
| **Applicable for large-scale, complex, or innovative designs:** | Yes | Not normally* |
| **Design option:** | No | Yes |
| **Typical project value range:** | $25K or greater | Up to low millions** |
| **Basis for award:** | Competitive Sealed Bids or Proposals | RFP, RFQ, or RFQ/RFP |
| **Base contract pricing structure:** | Contractor-established Stipulated Sum | Pre-established Unit Price x Coefficient*** |
| **Pricing structure for owner changes:** | Subcontractor-based + Contractor fee | Pre-established Unit Price x Coefficient*** |
| **Potential for lengthy DBB-type pre-construction time lines. (Refer to Figure 1.1.):** | Yes | No |
| **Contractor technical assistance & value engineering during design phase:** | No | Yes |
| **Fosters teamwork & partnering:** | No | Yes |
| **Approximate risk to owner:** | Low    Medium    High | Low    Medium    High |

**Figure 1.3** shows how the features of the stipulated sum project delivery method compare with those of JOC. Note that there are few similarities.

\* Sometimes the partnership inherent in JOC enables the method for use with complex and/or innovative design, or when unusual requirements require tight coordination.

\*\* Less for new construction; more for renovations. JOC projects can run from the low hundreds up to the low millions of dollars—but most are at the low end. (Federal limit is $3M.)

\*\*\* Other modifiers may sometimes exist and are applied to the sum total of the pre-established unit price × coefficient. (Contractual requirements may vary.)

## *Bid Documents*

The time frame for preparing and submitting bids is set by the owner in accordance with applicable governing requirements. At the designated time and location, bids are publicly read aloud. The contract award is based on the lowest bid from the contractor who meets all the minimum qualifications criteria set forth in the bid documents, such as bonding capability, proof of required insurances, and, if applicable, a *historically underutilized business (HUB)* plan. After awarding the project, the bid documents are established as the basis for the contract and are referred to from that time forward as the *contract documents.*

If the high bidder is the only one to meet the qualifications criteria set forth in the bid documents, and there is not enough time to repeat the solicitation process, the high bidder may be awarded the contract (although generally without the owner's enthusiasm). In this case, one alternative would entail rejecting all bids as being out of the established budget range, which would most likely postpone the project. Although not an everyday practice in the industry, this action would allow time for the owner to re-examine the project bid documents in regard to cost-saving alternatives and adjust the project's performance date to obtain pricing within the established budget. Another alternative would be to award the contract to the lowest qualified bidder and discuss ways to adjust the scope of work to fit the established budget, if this option is feasible in relation to the project and sanctioned by the owner's procurement requirements.

If applicable, the owner may consult with the procurement staff having jurisdiction, which may result in ways to enlarge the bidding pool. This usually enhances the number of bidders, and therefore the likelihood of healthy competition.

## Advantages & Disadvantages

Competitive sealed bid contracts pose a certain amount of risk to the owner. Low-bidding contractors often focus mainly on the bottom line, which may hamper the quality of work performed. This can put additional pressure on the owner and/or designer to ensure quality workmanship. The challenge in this situation can be somewhat mitigated if the awarded contractor has undergone a formal prequalification process. Frequently, owners will prequalify contractors, solicit bids from only that particular pool of contractors, and then base the award on the lowest bid, which helps to ensure that a qualified contractor is selected. Owner construction contract award requirements may vary pertaining to contractor prequalification practices.

Stipulated sum competitively bid contracts can, in some cases, offer the owner significant savings. For example, in a healthy competitive bidding situation, contractors might submit bids with lower-than-usual profit margins to keep their labor resources intact. On the other hand, these types of contracts can offer a substantial profit margin to the contractor in cases where there is a lack of healthy competition and a bid submitted with a higher profit margin has a good chance of being the lowest bid received by the owner.

## Competitive Sealed Proposals— Stipulated Sum

*Competitive Sealed Proposals—Stipulated Sum* delivery methods are similar to competitively bid stipulated sum contracts in their development of bid documents, the agreement between the owner and the contractor to work for a stipulated amount, and the solicitation process. The primary difference is that the proposal submitted by the contractor undergoes an evaluation process based on additional criteria to the dollar amount. Other factors affecting award usually include the contractor's capabilities, together with a proposed work plan for completing the project. The best overall value to the owner for contractor selection is identified using a point system whereby the contractor receiving the highest individual score wins the award.

After award and within governing guidelines, competitive sealed proposals sometimes allow the owner to modify the project scope and engage in associated negotiations with the contractor. A predetermined method for allowing modifications must be addressed in the bid documents, and usually are based on contractor-established unit costs, which are bid as alternates at the time the proposal is submitted.

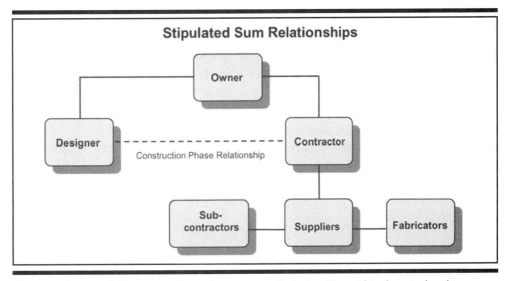

**Figure 1.4** uses solid lines to indicate the contractual relationships within the stipulated sum delivery method. The designer/contractor relationship, if it exists, is construction-phase only and process-oriented (dashed line). The contractor may use an in-house workforce in lieu of, or in addition to, subcontractors.

# Cost Plus Fee

A *Cost Plus Fee (CPF)* delivery method compensates the contractor for total project expenditures accrued for the work, plus a stipulated fee (or fee based on a percentage of the total project cost). The CPF contractor has similar responsibilities to those of stipulated sum contract, but without the associated financial risks. The contractor still schedules, coordinates, and manages the work, including overseeing subcontractor bidding and engaging in all construction contracts required for the project. As determined by contract provisions, a fee is added for the contractor's services rendered, overhead, and profit. By this method, the owner contracts separately with a designer.

CPF delivery methods are project-specific and used mostly in the private sector for smaller projects. CPF can include innovative or

## Comparison of Features: Cost Plus Fee vs. JOC

| Delivery Method Features | Delivery Method | |
|---|---|---|
| | **Cost Plus Fee** | **JOC** |
| Contract type:<br>Term of contract:<br>Primary application: | Project-specific<br>Project-specific<br>Multi-trade | ID/IQ<br>1 year + options (typical)<br>Multi-trade |
| Applicable for large-scale, complex, or innovative designs | Not normally large-scale designs, unless schedule requires concurrent designing and building | Not normally* |
| Design option: | No | Yes |
| Typical project value range: | $5K to $25K | Up to low millions ** |
| Basis for award: | Informal or Formal Competitive Bids | RFP, RFQ, or RFQ/RFP |
| Base contract pricing structure: | Actual project costs + Contractor-established fee | Pre-established Unit Price x Coefficient*** |
| Pricing structure for owner changes: | Subcontractor-based + Contractor fee | Pre-established Unit Price x Coefficient*** |
| Potential for lengthy DBB-type pre-construction time lines.<br>(Refer to Figure 1.1.): | Reduced but possible | No |
| Contractor technical assistance & value engineering during design phase: | No | Yes |
| Fosters teamwork & partnering: | No | Yes |
| Approximate risk to owner: | | |
| | Low       Medium       High | Low       Medium       High |

**Figure 1.5** compares the characteristics of a CPF delivery method with those of JOC. Note that there are few similarities. One major difference is the approximate risk to the owner with CPF, since there is no fixed price for the project established at any time before or after the contract award.

* Sometimes the partnership inherent in JOC enables the method for use with complex and/or innovative design, or when unusual requirements require tight coordination.

** Less for new construction; more for renovations. JOC projects run from the low hundreds up to the low millions of dollars—but most are at the low end. (Federal limit is $3M.)

*** Other modifiers may sometimes exist and are applied to the sum total of the pre-established unit price × coefficient. (Contractual requirements may vary.)

complex design arrangements when quick project starts are desired but the design has not yet been completed.

With CPF, even if design documents are complete, accurate cost estimating can be difficult to achieve. However, construction can begin after contract award and prior to the completion of design documents. Financial risk may be decreased for the owner if the contractor's fee is set for the project as a whole, instead of as a percentage of total costs. Risk is always increased for both the contractor and subcontractor when stipulated amounts are bid for their services. The main disadvantage for the owner using the CPF method is that total project costs are unknown before the construction phase begins—and difficult to determine until the project nears completion.

## Cost Plus Fee with Guaranteed Maximum Price (CPF-GMP)

This delivery method is similar to CPF, except that it has a cost-capping feature that reduces the owner's risk. CPF-GMP is more popular than CPF for projects of higher value. Figure 1.6 compares the features of CPF-GMP with those of JOC.

Since construction projects vary in scope and complexity, they can challenge even the most adept designers to produce 100% complete design documents within a constrained time frame. Often, the owner's budget dictates a maximum expenditure on the project design effort, precluding the attention to detail normally present in stipulated sum contracts. If a project must begin before the design documents are complete, a CPF-GMP contract is one way to minimize the associated risk of cost uncertainties to the owner.

The CPF-GMP delivery method is similar to the CPF method. The exception is that certain provisions enable the bidder to project a maximum construction project cost, along with a set fee for overhead and profit. The design documents need to be concise enough for the contractor to estimate all the projected construction costs with comfortable precision. The more design detail, the better, for the owner. Otherwise, the contractor could increase contingencies for work that lacks detail and embed them into the guaranteed maximum price.

With CPF-GMP contracts, the contractor absorbs any overrun of the maximum price within the established scope of work. The contractor, owner, and/or designer mutually agree on any changes in the scope of work, with the acknowledgment that these changes will either increase or decrease the maximum cost of the project. Cost estimates associated with pending changes should be as accurate as possible for careful consideration of their impact on the project budget. CPF-GMP contracts can include provisions for the owner and the contractor to share either project cost savings or cost overruns via a sharing clause. It is a current practice to include the GMP as a provision within other contract types, such as Construction Manager-at-Risk (CMAR). Figure 1.7 illustrates CPF relationships.

## Comparison of Features: Cost Plus Fee-GMP vs. JOC

| Delivery Method Features | Delivery Method | |
|---|---|---|
| | **CPF-GMP** | **JOC** |
| Contract type: | Project-specific | ID/IQ |
| Term of contract: | Project-specific | 1 year + options (typical) |
| Primary application: | Multi-trade | Multi-trade |
| Applicable for large-scale, complex, or innovative designs: | Possible | Not normally* |
| Design option: | No | Yes |
| Typical project value range: | $25K or greater | Up to low millions** |
| Basis for award: | Informal or Formal Competitive Bids | RFP, RFQ, or RFQ/RFP |
| Base contract pricing structure: | Actual project costs + Contractor-established fee | Pre-established Unit Price x Coefficient*** |
| Pricing structure for owner changes: | Subcontractor-based + Contractor fee | Pre-established Unit Price x Coefficient*** |
| Potential for lengthy DBB-type pre-construction time lines. (Refer to Figure 1.1.): | Reduced but possible | No |
| Contractor technical assistance & value engineering during design phase: | No | Yes |
| Fosters teamwork & partnering: | No | Yes |
| Approximate risk to owner: | Low    Medium    High | Low    Medium    High |

**Figure 1.6** compares the features of a CPF-GMP delivery method with those of JOC. Note that there are few similarities, with the main differences being pricing structure (primarily based on actual costs) and the lack of contractor technical assistance during the design phase.
* Sometimes the partnership inherent in JOC enables the method for use with complex and/or innovative design, or when unusual requirements require tight coordination.
** Less for new construction; more for renovations. JOC projects run from the low hundreds up to the low millions of dollars—but most are at the low end. (Federal limit is $3M.)
*** Other modifiers may sometimes exist and are applied to the sum total of the pre-established unit price x coefficient. (Contractual requirements may vary.)

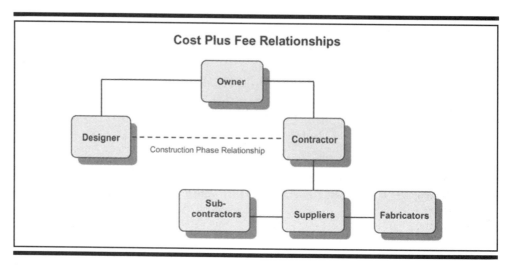

**Figure 1.7** illustrates, using solid connecting lines, the contractual relationships in CPF and CPF-GMP delivery methods. The designer/contractor relationship (dashed line), if it exists, does so during the construction phase only and is process-oriented. The contractor may use an in-house workforce in lieu of, or in addition to, subcontractors.

# Construction Manager

A *Construction Manager (CM)* or *Construction Management Agency (CMA)* provides professional construction management and technical services to the owner for a fee. CMs serve as consultants to the owner, typically providing design phase assistance; technical expertise; estimating, scheduling, and project coordination; and inspection of the work as it progresses. Figure 1.8 compares the characteristic features of a CM to those of JOC.

CMs also provide value engineering, bid coordination, and other construction management services. They usually do not have direct contractual relationships with designers, contractors, subcontractors, fabricators, or materials vendors and are not fiscally responsible for these project expenditures. CMs can assist the owner in site assessments and establishing and controlling the scope of the work for the project. They can also provide advice on the optimum use of funds

### Comparison of Features: Construction Manager vs. JOC

| Delivery Method Features | Delivery Method | |
|---|---|---|
| | **Construction Manager** | **JOC** |
| Contract type: | Project-specific | ID/IQ |
| Term of contract: | Project-specific | 1 year + options (typical) |
| Primary application: | Multi-trade | Multi-trade |
| Applicable for large-scale, complex, or innovative designs: | Yes | Not normally* |
| Design option: | No | Yes |
| Typical project value range: | $25K or greater | Up to low millions** |
| Basis for award: | Request for Qualifications | RFP, RFQ, or RFQ/RFP |
| Base contract pricing structure: | Actual project costs + CM negotiated fee | Pre-established Unit Price x Coefficient*** |
| Pricing structure for owner changes: | Subcontractor-based + CM fee | Pre-established Unit Price x Coefficient*** |
| Potential for lengthy DBB-type pre-construction time lines. (Refer to Figure 1.1.): | Yes | No |
| CM technical assistance & value engineering during design phase: | Yes | Yes |
| Fosters teamwork & partnering: | Yes | Yes |
| Approximate risk to owner: | Low   Medium   High | Low   Medium   High |

**Figure 1.8** compares the CM or CMA delivery method with JOC. Similarities include technical assistance provided during the design phase and fostering teamwork among project participants.
\* Sometimes the partnership inherent in JOC enables the method for use with complex and/or innovative design, or when unusual requirements require tight coordination.
\*\* Less for new construction; more for renovations. JOC projects run from the low hundreds up to the low millions of dollars—but most are at the low end. (Federal limit is $3M.)
\*\*\* Other modifiers may sometimes exist and are applied to the sum total of the pre-established unit price x coefficient. (Contractual requirements may vary.)

budgeted for the construction process. The award criterion is primarily the CM's qualifications, with consideration of a proposed or negotiated fee.

CM services can be advantageous to owners, especially those with limited internal project administration experience in certain areas, including the following:

- Keeping the budget on track
- Meeting desired project time lines and avoiding delays
- Evaluating and approving changes to the scope of work
- Monitoring quality control
- Resolving disputes

The CM also provides third-party impartiality by working directly in the interest of the owner. However, these benefits must be weighed against the probability that the owner will engage in multiple contracts and deal directly with subcontractors, fabricators, material vendors, and others. In this scenario, there is the likelihood that the owner may not be able to transfer various risks to a single point of responsibility, as with other project delivery methods.

If the owner contracts with a prime or general contractor only, some of the CM's services may already be provided by the designer or contractor in accordance with traditional provisions of designer-owner and owner-contractor agreements. Nonetheless, using CMs on medium to large capital and renovation projects has been documented as a successful means of project delivery.

## Construction Manager-at-Risk

The *Construction Manager-at-Risk* (CMAR) method of project delivery is currently growing in popularity and gathering endorsement among public-sector entities as being well-suited for medium to large capital or renovation projects. The success of CMAR is most directly attributable to the following:

- Its provision of technical assistance to the designer during the design phase
- Its GMP cost-capping feature
- The ability to start construction before design documents are 100% complete

This contract type is similar to CPF-GMP, with the addition of services typically provided to the owner under a CM contract. The "at-risk" identifier denotes a single point of responsibility for the entire construction process. Figure 1.11 compares the characteristic features of the CMAR method with those of JOC.

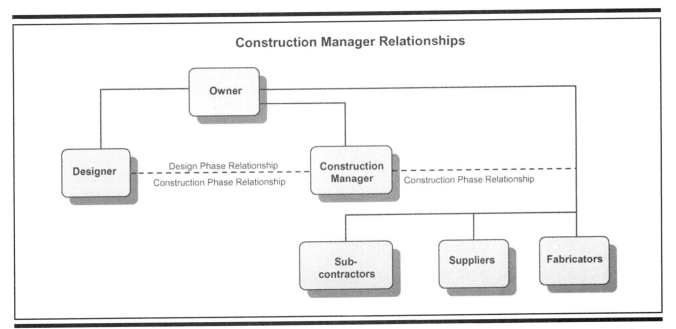

Figure 1.9 illustrates the contractual relationships in CM delivery methods—indicated by solid lines. There is no contractual relationship between the designer and the CM, nor are there contractual relationships between the CM and subcontracted resources. The designer/CM relationship (dashed line) includes design and construction phase assistance. The CM provides assistance to the owner during the construction phase (dashed line).

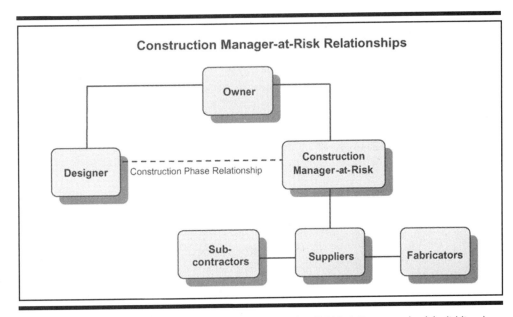

Figure 1.10 illustrates the contractual relationships in the CMAR delivery method (solid lines). The designer/CMAR relationship includes design and construction phase assistance (dashed line). The CMAR may use an in-house workforce in addition to subcontractors.

## Comparison of Features: Construction Manager-at-Risk vs. JOC

| Delivery Method Features | Delivery Method | |
|---|---|---|
| | CMAR | JOC |
| Contract type: | Project-specific | ID/IQ |
| Term of contract: | Project-specific | 1 year + options (typical) |
| Primary application: | Multi-trade | Multi-trade |
| Applicable for large-scale, complex, or innovative designs: | Yes | Not normally* |
| Design option: | No | Yes |
| Typical project value range: | $25K or greater | Up to low millions** |
| Basis for award: | RFQ or RFQ/RFP | RFP, RFQ, or RFQ/RFP |
| Base contract pricing structure: | GMP inclusive of CMAR negotiated fee | Pre-established Unit Price x Coefficient*** |
| Pricing structure for owner changes: | Subcontractor-based + CMAR fee | Pre-established Unit Price x Coefficient*** |
| Potential for lengthy DBB-type pre-construction time lines. (Refer to Figure 1.1.): | Reduced | No |
| CMAR technical assistance & value engineering during design phase: | Yes | Yes |
| Fosters teamwork & partnering: | Yes | Yes |
| Approximate risk to owner: | Low    Medium    High | Low    Medium    High |

**Figure 1.11** shows how CMAR compares with JOC. Both methods provide technical assistance during the design phase, foster teamwork among project participants, and offer cost-capping. The primary difference is in their pricing structures. (CMAR is project-specific and well-matched to large projects.)

\* Sometimes the partnership inherent in JOC enables the method for use with complex and/or innovative design, or when unusual requirements require tight coordination.

\*\* Less for new construction; more for renovations. JOC projects run from the low hundreds up to the low millions of dollars—but most are at the low end (federal limit is $3M).

\*\*\* Other modifiers may sometimes exist and are applied to the sum total of the pre-established unit price x coefficient. (Contractual requirements may vary.)

## Bidding

The CMAR award process for the public sector is similar to design-build, beginning with solicitations to firms for RFQs. From that pool, a *short-list* of three firms is typically selected for interview, and then ranked in order of preference. Often there is little or no design criteria developed at the time the RFQ is posted. The award is based on the following:

- A thorough evaluation of the CMAR's qualifications
- The firm's lump-sum or negotiated service fee
- The CMAR's ability to assist in the design phase (including the development of a GMP)
- A proposed time line for the project

If the fee for construction services cannot be negotiated with the first candidate, the next candidate receives an opportunity to engage in negotiations, and so forth, until a firm is chosen or the RFQ solicitation is advertised again. In considering the fee, it is important to clearly define what will be covered in the General Conditions, as this directly impacts the actual amount received by the CMAR.

## Design Phase Assistance

With CMAR, the owner establishes a separate contractual relationship with the designer. This arrangement allows for unbiased impartiality and promotes checks and balances of the construction process on the owner's behalf. The designer is made aware of the CMAR's design phase assistance requirements and is expected to partner in the process.

The owner contracts with the CMAR for total construction phase services and design phase assistance, including cost estimating, scheduling, technical expertise, value engineering efforts, and developing a GMP when design documents near completion. The CMAR, designer, and owner establish a project team to coordinate pre-construction efforts. As soon as is practical during the design phase, the GMP for the project is developed with the CMAR in agreement to pay all costs that exceed this amount, unless changes are authorized by the owner.

## Project Management

The CMAR is responsible for conducting the subcontractor bidding process, sometimes prequalifying subcontractors with approval of the designer and/or owner. The CMAR contracts directly with subcontractors, suppliers, and fabricators; and schedules, coordinates, and manages the construction effort. Construction can commence in phases as soon as is practical after the GMP has been set and those portions of the design documents are completed. The set fee for services allows the CMAR to stay focused on obtaining and scheduling project resources instead of being concerned about profit margins during the course of construction.

## Design-Build

*Design-build* has gained popularity in the private sector, in part due to its potential for accelerated pre-construction time lines, which can help owners offset interim construction loan interest rates. Legislative changes in both federal and state procurement regulations have allowed design-build to be used increasingly for public-sector projects. While the traditional public-sector contractor award process is based on the bid amount, designer selection traditionally has been based on qualifications, with fees negotiated after selection.

Design-build allows an owner to contract with a single entity for both design and construction services. This feature is similar to JOC when a larger, specialized JOC contractor has professional A/E consultants under contract or direct employment. The method is well-matched for both large, complex projects, such as refineries and pharmaceutical plants, and simple office buildings and warehouses. Figure 1.12 compares the features of design-build with those of JOC.

## Bidding

In the public sector, the award process typically involves open solicitation of qualification statements from design-build firms. A prequalified short-list of respondents is then determined. The selected few are interviewed by committees as they submit price proposals for

---

### Comparison of Features: Design-Build vs. JOC

| Delivery Method Features | Delivery Method | |
| --- | --- | --- |
| | Design-Build | JOC |
| Contract type: | Project-specific | ID/IQ |
| Term of contract: | Project-specific | 1 year + options (typical) |
| Primary application: | Multi-trade | Multi-trade |
| Applicable for large-scale, complex, or innovative designs: | Yes | Not normally* |
| Design option: | Yes | Yes |
| Typical project value range: | $25K or greater | Up to low millions** |
| Basis for award: | RFQ/RFP | RFP, RFQ, or RFQ/RFP |
| Base contract pricing structure: | GMP + fee | Pre-established Unit Price x Coefficient*** |
| Pricing structure for owner changes: | Subcontractor-based + DB fee | Pre-established Unit Price x Coefficient*** |
| Potential for lengthy DBB-type pre-construction time lines. (Refer to Figure 1.1.): | Reduced | No |
| DB technical assistance & value engineering during design phase: | Yes | Yes |
| Fosters teamwork & partnering: | Yes | Yes |
| Approximate risk to owner: | Low    Medium    High | Low    Medium    High |

**Figure 1.12** shows the similarities of design-build and JOC in terms of technical assistance provided by the contractor during the design phase. This approach fosters teamwork among project participants and also promotes cost-capping. The primary differences include pricing structure and the project-specific nature of design-build.

* Sometimes the partnership inherent in JOC enables the method for use with complex and/or innovative design, or when unusual requirements require tight coordination.

** Less for new construction; more for renovations. JOC projects run from the low hundreds up to the low millions of dollars—but most are at the low end. (Federal limit is $3M.)

*** Other modifiers may sometimes exist and are applied to the sum total of the pre-established unit price × coefficient. (Contractual requirements may vary.)

the project under consideration, usually at a guaranteed maximum price with set fees. Often, there is little or no design criteria developed at the time of the RFQ posting. Sometimes the design-build firm's fee is considered together with the firm's projected time line for the project, referred to as "A+B" bidding within a design-build proposal.

## Advantages & Disadvantages

There are a number of advantages for owners using a single point of contact, as is the case with design-build. There is also the benefit of the partnering principles inherent within this project delivery method due to the merging of two disciplines. With the contractor and designer working together, estimating accuracy increases, because costs can be projected while the design is being developed. Design-build also allows for earlier project starts, because initial stages of construction can proceed while the final design for subsequent phases is being completed.

When little or no design criteria has been developed at the time of the contract award, a successful design-build project will carefully define the design criteria by establishing those design parameters that meet the user group's needs and desires during programming. Open collaboration between the design-build firm and focus groups of potential users is essential in establishing satisfactory end results in design criteria development. Failure to sufficiently include user focus group input may result in a facility that is not well-suited for its desired purpose.

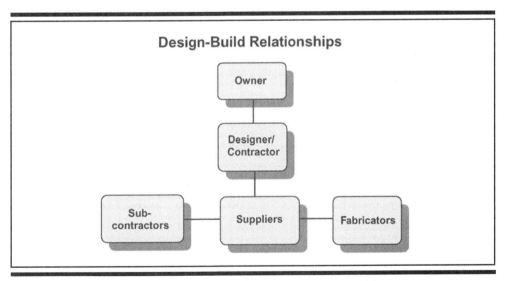

**Figure 1.13** uses solid lines to illustrate design-build contractual relationships. The designer/builder can be a single firm, contractor-led, designer-led, a joint venture, or subcontracted. The design-build contractor may use an in-house workforce in addition to subcontractors.

The design-build firm is responsible for submitting practical solutions to the owner for conflict resolution and/or omissions to stay within the project budget. Problems can be identified and resolved during the preliminary design phase rather than after the design is completed. Most design-build firms today are contractor-led (instead of designer-led), since construction companies historically have more sophisticated scheduling and cost control techniques. Whichever discipline leads the firm, the design professional in the team can be faced with increased risks and liabilities not typically seen with independent design contracts. This is because as affiliates with design-build firms, design professionals can be responsible for construction contracting liabilities in addition to design liabilities.

The contractor-designer partnership may be of concern to some owners, because the designer may not be as impartial or objective in serving the owner's best interests. Even so, design-build is enjoying widespread popularity due to the flexibility it provides as it continues to evolve. The method can contain contract provisions to include integration of the owner's workforces when desired, building operations management and maintenance after construction, and project financing options.

## Time & Materials

The *Time & Materials (T&M)* delivery method allows for contractors to be compensated based on the total hourly labor rate bid applied to actual work time expended, plus materials purchased and equipment used. The method is best for small projects under certain circumstances, such as those requiring little or no design, together with the need for rapid mobilization. The T&M method can be similar to the CPF method when documenting actual construction costs and adding contractor service fees. However, sometimes T&M labor rates include the contractor's overhead and profit (O&P) within each individual classification of in-house workforce (project supervisor, craftsperson, apprentice, laborer, etc.). Additionally, CPF methods are usually project-specific, and T&M frequently utilizes the ID/IQ option. T&M is usually not recommended for larger projects due to the pricing structures based on the unknown time that will be accrued to perform the tasks, which promotes uncertainty in the projection of overall project costs.

The T&M method can be used for tasks involving single trades or grouped trades as a way to handle changes to the scope of work within project-specific, lump-sum methods, if allowed by the contract. To validate contractor activity, the owner should provide a high level of visibility at the project site. This requires an owner's representative, usually an inspector or contract administrator, to provide significant oversight of contractor activities due to the potential for fluctuations in the level of contractor labor productivity. These labor costs are absorbed by the owner as the project progresses, as shown in the following example:

> *A worker makes an error in material layout on a project. The problem is discovered only after the materials have been cut and installation has begun. With intent to correct the situation, the worker quickly removes and discards the materials already applied. The layout is corrected, and new material is cut and correctly installed. This happens in a short interval during the absence of the owner's resident project representative.*

In the above example, under T&M contract provisions, it is possible for the owner to unknowingly absorb the cost for labor on the original layout, removal of the erroneously installed materials, and the original materials (which ended up in the project's dumpster).

---

### Comparison of Features: Time & Materials vs. JOC

| Delivery Method Features | Delivery Method | |
| --- | --- | --- |
| | **Time & Materials** | **JOC** |
| Contract type: | Project-specific or ID/IQ | ID/IQ |
| Term of contract: | Project-specific or 1 year | 1 year + options (typical) |
| Primary application: | Single or grouped trades | Multi-trade |
| Applicable for large-scale, complex, or innovative designs: | No | Not normally* |
| Design option: | No | Yes |
| Typical project value range: | Up to $25K | Up to low millions** |
| Basis for award: | Competitive bids | RFP, RFQ, or RFQ/RFP |
| Base contract pricing structure: | Total hours expended + materials with fee | Pre-established Unit Price x Coefficient*** |
| Pricing structure for owner changes: | Total hours expended + materials with fee | Pre-established Unit Price x Coefficient*** |
| Potential for lengthy DBB-type pre-construction time lines. (Refer to Figure 1.1.): | If project-specific: Rare If ID/IQ: No | No |
| Contractor technical assistance & value engineering during design phase: | No | Yes |
| Fosters teamwork & partnering: | No | Yes |
| Approximate risk to owner: | Low  Medium  High | Low  Medium  High |

Figure 1.14 compares T&M with JOC. It shows that T&M can have ID/IQ features and that both methods can be used for small project delivery as well as rapid mobilization. The owner's financial risk can be high with T&M.

* Sometimes the partnership inherent in JOC enables the method for use with complex and/or innovative design, or when unusual requirements require tight coordination.

** Less for new construction; more for renovations. JOC projects run from the low hundreds up to the low millions of dollars—but most are at the low end. (Federal limit is $3M.)

*** Other modifiers may sometimes exist and are applied to the sum total of the pre-established unit price × coefficient. (Contractual requirements may vary.)

From an owner's standpoint, T&M contracts are best suited to trade projects or tasks. T&M can also be a good contracting tool on smaller projects when quick mobilization by the contractor is desired and total project costs are not an issue of major concern for the owner.

## Bidding

For public-sector owners, the T&M contractor selection process may also be challenging, since classified labor rates tend to fluctuate. In this case, contractor evaluation and selection using "best overall value" criteria can be applied. Previously completed projects can also be evaluated for actual labor expenditures. Labor rates proposed by each bidder can be applied categorically to those projects for comparison. Contract provisions should require contractor compliance with established minimum labor rates for various trades, as applicable.

If historic project cost data is not available to the owner, construction cost estimating reference texts, such as Means *Building Construction Cost Data,* can be a valuable resource for estimating task-specific labor-hours of workers. In any case, it is always best for the owner to collaborate with the procurement entity having jurisdiction to determine the contractor award.

## Documentation

T&M contracts require contractors to generate and maintain accurate documentation of project costs as the work progresses. The contractor's payroll, materials invoicing, and equipment rental should be carefully reviewed. Some contracts may require third-party certification prior to payment. Contractor compensation can be based on the following:

- Documentation of the aggregate hours expended on the project per classification of labor forces
- Invoices for materials, equipment rental, and fabrication of specified specialty items
- Expenditures for mobilizing, maintaining, and demobilizing the contractor's field office, if applicable
- Associated fees for materials and project management, as applicable

Documenting all reimbursable resources associated with a T&M project can be burdensome to all parties involved, especially in terms of overhead and profit margins, unless the scope is relatively small.

Most contractors and subcontractors favor T&M contracts, since they offer the benefit of virtually no financial risk. On the other hand, if the project does not have a projected end date, it may be difficult for the contractor to schedule other future work.

# *Unit Price*

This delivery method is typically used when work is of a repetitive nature or when quantities of work required are unknown (such as some site work projects) and performed by a single trade or a group of closely related trades. Work is limited to the tasks identified by the owner for a specific project or a series of projects with similar tasks. This method is commonly used by facility owners to augment in-house trades during peak periods. Often, trade subcontractors use established unit price-based pricing structures when submitting bids to general contractors.

Projects can be divided into listings of specific, descriptive task components, qualified as individual task *line items* of the work. Each line item consists of labor, material, and sometimes equipment data expressed in *units*. Each unit of work can then be subsequently unit priced, thereby designating a cost per unit *(unit cost)* for a specific task performed. The methodology of unit pricing a project entails calculating the sum total of unit prices, based on the quantity of units for each qualified line item.

Unit price-based project delivery methods enable compensation to the contractor based on either the measurement of units of actual *work-in-place,* or the units of work estimated to be performed, with the estimate serving as a fixed price or lump sum. When compensation is based on measurement of work-in-place, completed line items of work are identified *(qualified),* and the associated units are measured or counted *(quantified).* With unit-price delivery methods, unit prices are established by the contractor's bid.

## Bidding by Labor Units

Sometimes established ID/IQ unit pricing does not involve materials, and work is defined in *labor units* per hour. In this case, the labor unit price may include a qualified tradesperson, with or without an apprentice or helper, having a job truck fully outfitted with the tools of the trade. Materials can be furnished by the owner, and the worker(s) act(s) as a possible extension of a facility owner's in-house workforce, as needed. To reduce risks, the owner's representative (typically a trade supervisor) and the ID/IQ contractor jointly estimate the number of labor unit hours anticipated to complete a delegated task on a not-to-exceed basis.

Compensation is calculated on actual labor unit hours expended up to the estimated amount, with overages of time absorbed by the ID/IQ contractor. Unit-price delivery methods that exhibit ID/IQ features typically have terms of up to one year. As pricing structures are established by the contractor, the method typically does not allow for fluctuations in actual costs to the contractor due to labor rate or materials cost increases.

## Comparison of Features: Unit-Price Contracts vs. JOC

| Delivery Method Features | Delivery Method | |
|---|---|---|
| | Unit-Price Contracts | JOC |
| Contract type: | Project-specific or ID/IQ | ID/IQ |
| Term of contract: | Project-specific or 1 year | 1 year + options (typical) |
| Primary application: | Single or grouped trades | Multi-trade |
| Applicable for large-scale, complex, or innovative designs: | No | Not normally* |
| Design option: | No | Yes |
| Typical project value range: | Usually up to $25K | Up to low millions** |
| Basis for award: | Competitive bid | RFP, RFQ, or RFQ/RFP |
| Base contract pricing structure: | Contractor-established unit price | Pre-established Unit Price x Coefficient*** |
| Pricing structure for owner changes: | Contractor established unit price and/or T&M | Pre-established Unit Price x Coefficient*** |
| Potential for lengthy DBB-type pre-construction time lines. (Refer to Figure 1.1.): | If project-specific: Possible If ID/IQ: No | No |
| UP technical assistance & value engineering during design phase: | No | Yes |
| Fosters teamwork & partnering: | No | Yes |
| Approximate risk to owner: | Low    Medium    High | Low    Medium    High |

**Figure 1.15** compares unit-price delivery methods to those of JOC. While some unit-price methods have ID/IQ features, both methods can be used on small projects and are capable of rapid mobilization, as needed, for quick project starts.

\* Sometimes the partnership inherent in JOC enables the method for use with complex and/or innovative design, or when unusual requirements require tight coordination.

\*\* Less for new construction; more for renovations. JOC projects run from the low hundreds up to the low millions of dollars—but most are at the low end. (Federal limit is $3M.)

\*\*\* Other modifiers may sometimes exist and are applied to the sum total of the pre-established unit price × coefficient. (Contractual requirements may vary.)

## Bidding by Task Units

Unit-pricing can be established by the contractor for each component task of the overall work identified by the owner in the bid documents. The sum total of unit prices proposed by the contractor for the identified tasks becomes an evaluation factor for contract award, but serves only as a guideline for compensation.

The owner-estimated quantities of tasks provide the contractor with information needed to identify the project's required resources. This helps the contractor develop a bid for the work. As work progresses, compensation to the contractor is not based on the owner's estimated quantities, but rather on field measurement of the exact quantities of work-in-place, as applied to the unit-price bid for each respective task performed.

The following example describes an owner using a rough estimate of work defined in bid documents as task units. These task units serve as a guideline for bidder pricing, with the actual quantities of work-in-place used as the basis for payment.

*A city manager seeks to outsource construction of a roughly estimated quantity of sidewalks within municipality boundaries. The manager anticipates that not all sidewalks will be the same width. Exact locations and sequencing of the work are yet to be determined. Additionally, some of the sidewalks will include accessibility features, and signage will need to be relocated accordingly.*

*The contract documents include applicable standard design criteria and associated specifications for the work, along with a rough estimate of the overall length of the sidewalks to be constructed. Concrete contractors are solicited to provide unit-pricing in the form of cost per square foot. The accessibility features and signage are priced per unit. The contract term is for one year, with an option for renewal for an additional year. Compensation to the contractor is paid monthly, based on documentation of actual work performed in accordance with contract provisions.*

In this example, many of the site features of the work would need to be somewhat similar in order for the unit price offered to accommodate fair compensation to the contractor. Unit-price methods can also be applied to projects in which the resources needed are difficult to estimate—such as roads, parking lots, and other site-work-intense projects.

## Bid Evaluation

Unit-price bid documents that contain numerous line items for the contractor to bid may become difficult to evaluate, unless the bid documents reflect a specific project with fairly accurate owner-estimated quantities. This is because each task identified for bid will reflect each bidder's proposed price per unit for that specific task. The contract award is based on the most appropriate means of evaluation by the purchasing or procurement staff having jurisdiction, usually by application of proposed pricing to a representative project.

## Alternates

Unit-pricing, which sometimes is used in other project delivery methods as a basis of contractor compensation for changes to the scope of work, can also be applied to *alternates,* or items bid on in conjunction with the base bid of a specific project. Alternates are included in project-specific bid documents when an owner wants to obtain pricing for certain tasks or items as options that would modify the basic scope of work. Alternates might include *additive* or *deductive* tasks, or optional materials as substitutes for those that have been specified.

## Payment by Estimated Unit Prices

When contractor compensation is based on the unit prices for work to be performed, the cost estimate itself serves as a fixed-price or lump-sum payment criterion. In this case, the estimate must be as accurate as possible and typically is based on the project's written scope of work, inclusive of any related design documents. The cost estimate is mutually agreed to by both the owner and the contractor prior to the owner's issuance of authorization to perform the work. Unless changes in the scope of work are authorized by the owner, exact measurements of work-in-place may not be justification for additional compensation, depending on contract requirements. This unit price and payment feature would be similar to JOC if the unit prices were pre-established rather than established by the contractor as a response to the bid documents.

## Summary of the Unit-Price Method

Unit-price project delivery methods can be flexible and, when combined with ID/IQ features, can be useful as alternative delivery methods for small projects or tasks that require only single or grouped trades within the same CSI MasterFormat division. However, unit-price delivery can be problematic if the scope of work changes during construction to include tasks that were not prepriced by the contractor. If changes are expected, the owner may want to identify cash allowances or contingency funding as a means of compensation for the extra costs. Figure 1.16 illustrates unit-price relationships.

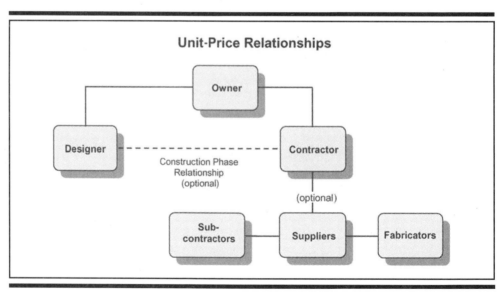

**Figure 1.16** uses solid lines to show unit-price contractual relationships. The contractor can have a construction-phase relationship with the designer if the owner has contracted for the service. The number of subcontractors, if any, would be limited. Additionally, the requirement for the contractor to furnish any materials or fabricated items is optional, depending on contract provisions. The contractor may use an in-house workforce in lieu of, or in addition to, subcontractors.

## Comparison of Delivery Methods by Project Values

Figure 1.17 shows various delivery methods that can be used for projects of different values. Some methods may fall into multiple value range categories. The comparison does not take into consideration projects of complex or innovative design, which may limit some methods from certain value range categories. This figure is intended only to provide a general comparative overview to assist owners in matching the most advantageous method with a project under consideration after a rough cost estimate for the project has been developed.

### Comparison of Delivery Methods by Project Values

| $0-$5K | $5K-$25K | $25K-Low $Ms | $10M and Over |
|---|---|---|---|
| • JOC<br>• Owner In-House Resources<br>• Single Source Contract<br>• T&M, Unit Price | • JOC<br>• Owner In-House Resources<br>• Cost Plus Fee<br>• Informal Competitive Bidding<br>• T&M, Unit Price | • JOC<br>• Competitive Sealed Bids or Proposals<br>• Cost Plus Fee-GMP<br>• Design-Build<br>• CMAR | • Competitive Sealed Bids or Proposals<br>• Cost Plus Fee-GMP<br>• Design-Build<br>• CM or CMA<br>• CMAR |

**Figure 1.17** shows the project delivery methods that typically are applied to projects of varying values. Note that JOC is well-matched for projects up to several million dollars, with the exception of major new construction and those with complex designs.

## Conclusion

Traditional project delivery methods usually involve lengthy contract award processes and pre-construction time lines. Traditional DBB methods have the potential for fluctuations in contractor performance and quality due to "low-bid" awards, resulting in increased contract management efforts by the owner.

The short-term relationships of project-specific delivery methods often do not foster opportunities to partner or give contractors sufficient time to become accustomed to the owner's requirements or expectations. "Process only" relationships—inherent with most other methods—increase the possibility of adversity and disputes among project participants. Other delivery methods provide contractors with less incentive to consistently provide high-quality customer service to the owner, unlike JOC, where there is the likelihood of future work contingent on high-quality, responsive work at a competitive price.

Most other methods reflect unstable pricing structures based directly on subcontractor pricing with the contractor's overhead and profit reflecting the specific project contract under consideration for award. In addition, the search for opportunities to initiate change orders and claims is an art form with some contractors. In fact, comprehensive guides exist for handling one of the most critical areas of construction contracting regarding profit maximization—how to discover all potential changes to construction projects. These guides include pricing, presenting, negotiating, and receiving payment for change orders. JOC does not foster this practice. JOC pricing structures do not allow inflated change order pricing traditionally experienced with most other delivery methods. Therefore, JOC contractors have no incentive to seek out change orders for profit maximization opportunities. Initiating claims can be detrimental to the long-term, mutually beneficial owner-contractor relationship representative of successful JOC contracts.

Construction and facilities professionals are taking notice of JOC's unique processes and advantages as compared to those of other project delivery methods. Results are achieved with:

- A strategic plan with emphasis on stakeholder "buy-in" and any needed training;
- A contract that meets the owner's needs and clearly defines all requirements, with emphasis on execution procedures and pricing structures;
- A thorough evaluation and award process to select a contractor who can offer the owner the best overall value;
- Consistent contract and project management, using clearly defined processes; and
- A partnering mindset among key participants.

When partnering is applied to clearly defined JOC processes, the combination can produce mutually beneficial results that continue and grow over time. This is, in part, attributable to JOC's potential for developing long-term relationships. The following chapters will explore each of these facets, along with the specific processes involved in job order contracting.

# Chapter 2
# JOC Fundamentals

## History of Job Order Contracting

Pre-established unit-cost project delivery methods have been in use for thousands of years. The Code of Hammurabi, written in the 18th century BC in what was then Babylonia, Mesopotamia, describes an arrangement whereby an owner pays a designated amount of money to a builder for each standard unit of measure applied to the surface area of a completed project. As noted in Chapter 1, today this method of project delivery is most commonly known as job order contracting, or JOC, in public and private sectors. (Some branches of the U.S. military have their own name for the method, as referenced in this section.) JOC was first adapted in the U.S. in the early 1980s by the military. Universities, federal agencies, K-12 educational facilities, state agencies, counties, and municipalities followed.

### Early 1980s

- In 1981, Major Harry Mellon, Chief Engineer for the U.S. contingent at NATO Headquarters in Brussels, adapted a local European method for contracting renovation and minor construction work into what is now known as job order contracting. This method was further adapted and used at the U.S. Army Military Academy at West Point. Impressed by the results, Colonel Bill Badger brought the success of this process to the attention of U.S. Army Chief of Engineers Lieutenant General Val Heiberg. Lieutenant General Heiberg was so impressed with the results and the process that he asked Major General Mark Sisnyiak, Chief of Military Programs, to guide the development of a proposal to run a pilot program. After many battles with Congressional and Small Business Administration staff, approval was finally gained for a pilot program.

### 1984–1985

- The U.S. Army established a pilot program of five one-year contracts to test JOC at Fort Ord, Fort Bragg, West Point, Fort Sam Houston, and Fort Sill. The U.S. Navy experimented with a small JOC contract at Mare Island (San Francisco).

### 1986–1987

- The U.S. Army, followed by the U.S. Navy, began full-scale use of JOC.
- The U.S. Air Force initiated similar contracts, referred to as Simplified Acquisition of Base Engineering Requirements (SABER).
- The U.S. Army Corps of Engineers provided and maintained unit-price books (UPBs) for all military branches using the method.

### 1989–1990

- Texas A&M and other universities began using JOC.

### 1991

- The U.S. Air Force introduced RSMeans cost data as the recommended standard for unit-price books at Castle Air Force Base.

### 1991–1992

- The National Institute of Health (NIH) and other federal agencies began using JOC.
- Spring Branch ISD (Houston) and other K-12 Texas schools began using JOC.

### 1992–2005

- JOC usage spread to include state agencies, counties, municipalities, and others. Since 2000 or so, JOC usage has expanded to include public-sector facilities among several states and continues to expand. JOC is included along with other services provided by state agency sanctioned purchasing cooperatives. RSMeans, the most recognized source of construction cost data in the U.S., began offering JOC-specific cost data in the form of a product called *Means JOCWorks*™.

Because JOC has a relatively short history in the public sector, there are a limited number of owners who are proficient in its use. Even so, the evolution is steadily progressing as participants receive training, realize successful implementation of their programs, and share their experiences with others. Proper practice of the method by industry professionals in conjunction with healthy competition among qualified contractors will ensure a bright future for JOC as participants quickly realize its inherent advantages.

# Understanding JOC Contracts

A *job order contract* is an indefinite delivery contract for indefinite quantities of construction over a specific duration of time. The contract includes both unit-price and indefinite delivery/indefinite quantity (ID/IQ) features. Contractors submit competitive sealed proposals (including their technical/management/performance capabilities) with price coefficients to be applied to the sum total of pre-established unit-priced tasks. These tasks are identified as *line items* and subsequently quantified as units of work for each project or job order (JO). Thus, a lump-sum, firm fixed price is established for each individual job order issued against the contract.

JOC contracts provide "on-call" contractor services to facility owners for projects that are yet to be determined on an as-needed basis. JOC is best suited for multiple minor construction, renovation, alteration, rehabilitation, and site work and landscape projects, as well as maintenance work. These types of projects are of relatively small scale and short duration. They often are required to be accomplished quickly, and sometimes in high volumes simultaneously—exactly what JOC is designed to deliver.

JOC has a unique set of contract terms, award processes, and project value requirements. These will be discussed in the following sections.

## Owner-Awarded, Cooperative-Awarded, & "Riders"

Job order contracts can be awarded either by individual facility owners or by purchasing cooperatives for use by their members. Of course, if a co-op JOC is not available in an owner's area, an owner-awarded JOC contract—or the use of another facility's contract through agreements—are the only other options.

### Owner-Awarded JOC Contracts

This type of contract gives the owner direct control over the development of contract requirements, as well as the RFP solicitation, evaluation, and award processes. The current trend for owners who desire a JOC contract is to implement their own JOC programs to match their individual requirements. Since JOC contractor profit margins are based on the volume of work they receive, there is a direct correlation between the potential for facility owners to obtain competitive, favorable pricing and the anticipated total annual value of *"JOCable"* work offered by the owner. In other words, the higher the anticipated value offered by the owner at the time of RFP solicitation, the better chance the owner has for cost savings per project in a healthy, competitive bidding atmosphere. This is a factor to be weighed when the anticipated annual project volume may not be enough for the owner to obtain a maximum potential for cost savings, or for some contractors to propose a competitive coefficient.

Owner-awarded JOC contracts have proven to be very successful for the owner and the contractor with the right contract, the right match in program participants, and with work authorized in alignment with (or in excess of) the amount anticipated at the time of contract award.

## Cooperative-Awarded JOC Contracts & "Riders"

Purchasing cooperatives are legally sanctioned by state codes or by purchasing regulatory agencies having jurisdiction over co-op sanctioning. Members of co-ops typically include public agencies, such as higher education facilities, school districts, municipalities, counties, state and federal agencies, and even private-sector facility owners. The more the membership grows, the greater the purchasing power the co-op has for offering commodities and services to its members, usually resulting in better value than can be realized through individual owners' purchasing power.

With cooperatives, areas (or regional boundaries) within a state are defined in compliance with an agreement with the sanctioning state governing body or agency. The cooperatives then enter into *inter-local* agreements (via membership) with facility owners within those predefined areas. Members are provided with a choice of commodities and services, which may include JOC services.

If the co-op offers JOC services, contracts are implemented through competitive sealed proposals in response to the co-op's JOC RFP. Awardees contract directly with the co-op to service their respective designated areas. A fee is paid to the co-op by the JOC contractor, or, in some cases, by the user-member of the co-op contract, based on a small percentage of each job order sum total amount awarded through the co-op contract. This fee compensates the co-op for costs associated with administering the contract and any marketing efforts.

JOC contracts awarded by purchasing cooperatives are an innovative way for facilities owners to have access to JOC contractors without having to go through an RFP solicitation and award process. This convenient option must be balanced against the value of having direct control over contractor selection, contract execution requirements, and contract pricing structures.

There has been much discussion among JOC practitioners about whether JOC services obtained through co-ops foster the same level of partnering among project participants as is typically present with successful owner-awarded contracts. In cases in which the co-op's contract requirements meet the needs of the member, and the JOC contractor providing service to the member performs well, JOC co-ops have proven successful. Long-term relationships can be fostered among the contractor, the owner, and even the co-op representative. To get the maximum benefit from the co-op relationship, the owner and the contractor should exploit the mutual advantages of their long-term relationship, similar to that of an owner-awarded contract. The contractor can then assist the owner in developing an overall program for JOC-type work and actually function as an unofficial extension of the owner's staff.

Co-op JOC services may be a particularly favorable option for owners with facilities in remote rural or widely spread metropolitan areas, as well as for owners with limited available contract management resources. Some cooperatives provide limited contract management consultation services to assist their members in executing contracts.

Cooperative-awarded JOC contracts may give facility owners discounts, depending on the project value size and volume of work authorized by the facility. Other benefits may include, in part, construction financing options. The advent of JOC services through cooperatives, or use of another facility owner's JOC contract through mutual agreement as a "rider" to the base contract, is relatively new. However, these methods are gaining popularity among facility owners as testimonials of successful results are shared.

## Job Orders

A *job order*, or JO, is a formal, written, project-specific authorization to accomplish work. It is issued to the JOC contractor by the owner during the term of the JOC contract. A JO reflects the acceptance of the contractor's proposal by the owner. JOs identify the following:

- The project and its design documents (if necessary)
- A detailed scope of work
- The project's performance times
- A lump-sum fixed price to be paid to the contactor as compensation for accomplishing the scope of work (this price reflects a detailed cost estimate approved by the owner)
- The owner's authorized representative(s)' signature(s) and date of issuance

Depending on the owner's purchasing procedures, a JO may constitute a prerequisite agreement to authorize work, subject to the issuance of a purchase order (PO).

## Unit-Pricing

A *unit-price book (UPB)* represents the cost data that serves as the basis for construction cost estimating. For JOC contracts, this data also establishes the base value for work to be accomplished in accordance with JOC contract provisions. Proficiency in the use of designated UPBs—by both the owner and the contractor—is essential for the success of a JOC contract.

The most widely used and accepted cost data among JOC practitioners are available from RSMeans. The Appendix of this book contains excerpts from RSMeans' *Facilities Construction Cost Data 2005,* which illustrate standardized cost data and will assist in understanding cost estimating terminologies and basic processes. These pages are useful for cross-referencing subsequent chapters of this book that address cost estimating and contract requirements. RSMeans has added a new product, *Means JOCWorks™,* which combines estimating and project management functions. *(See "Computer Software" in Chapter 3 and the Resources at the back of the book for more on* Means JOCWorks™.*)*

## Coefficients

The contractor's coefficient represents a proposed price multiplier in relation to the pricing structure defined within the JOC contract documents. This coefficient establishes a competitively bid cost adjustment, or multiplier, to UPB pricing, taking into consideration contract requirements in conjunction with the contractor's *anticipated* main office (and yet to be determined individual project) overhead and profit.

After award, the parties involved in the JOC contract utilize coefficients as multipliers to determine the value of each job order to be accomplished. The coefficients are applied to an individual project's sum total of qualified and quantified unit-priced tasks in relation to the following:

- The applicable wage rate category (prevailing, Davis-Bacon, etc.)
- The required performance time (standard/non-standard hours)
- Emergency work, if applicable

The proposal coefficient reflects a net decrease from or increase to either the "Bare Costs - Total" costs column or the "Total Incl. O&P" costs column in the designated UPB. While coefficients typically are carried to three decimal places, some contracts require only two decimal places (although some are carried to four places). It is important that decimal place limitations of coefficients, as well as the costs column to be used for coefficient application, be clearly referenced in the provisions of the contract.

The UPB line items and associated quantities reflect the tasks and prices to perform the scope of work for each individual job order. The sum total of unit prices is then multiplied by the contractor's coefficient. For example, if the Bare Costs - Total for painting 100 square feet of gypsum board is $23.00 and the contractor's applicable coefficient is 1.125, the adjusted unit price for performing the task will be $25.88 ($23.00 × 1.125 = $25.88).

Similarly, if the Bare Costs - Total sum of unit prices identified for a specific job order is $20,455.00, the adjusted cost of the project will be the lump-sum amount of $23,011.88 ($20,455.00 × 1.125 = $23,011.88). A contract provision that addresses the rounding of job order amounts to the nearest dollar might be considered. In this example, the lump-sum amount would then be $23,012. *(See "Price & Payment Procedures" in Chapter 5 as a cross-reference to this section regarding coefficient adjustments, including "city cost index" multipliers.)*

## Conclusion

Knowledge of JOC's origins and the fundamentals of how and why it is used can help readers understand its evolution into an effective project delivery method. Selection and designation of an appropriate UPB provides cost data critical to a JOC contract's pricing structure and takes the owner's needs and requirements into account. Proficiency in using the UPB, by all parties to the contract, is essential.

Basic terminology unique to JOC and identified in this chapter is commonly used among practitioners of the method and frequently referenced throughout subsequent chapters.

# Chapter 3

## JOC Program Implementation

Proper planning is critical to a JOC program's long-term success. With public-sector entities, the first step should be determining the applicability of JOC implementation in the facility owner's area of operations. This should take place even before strategic planning to identify goals, objectives, and necessary actions. Private-sector facility owners—although not restricted to delivery method options—are well-advised to understand JOC principles up front and properly plan for contract implementation.

The process of evaluating internal planning, design, construction resources, and purchasing processes, and then choosing the most effective program management methods to accommodate owner-contractor interface, lays the groundwork for successful contract execution. It also identifies the owner's training needs.

Most experienced JOC practitioners agree that owners new to the method should start out slowly, with perhaps only one or two small projects, to allow sufficient time for participants to become familiar with the process. The amount of effort needed to implement a JOC program is tied directly to the annual volume of anticipated work to be awarded by the method. Understanding basic JOC principles may be enough when using this method only intermittently. However, an owner will need a much higher level of familiarity with JOC contract management procedures for a program that has the potential to consistently award a high volume of work.

## Determining Applicability

The first step for public facilities owners considering a JOC program is to determine whether JOC is legally sanctioned for use in their locality. Because JOC is still relatively new in the industry, the method may not be specifically addressed by legislative law or public procurement regulatory bodies having jurisdiction over the owner's operations.

If not specifically addressed by state statute or other governing regulations, JOC may or may not conflict with established purchasing requirements for construction contract awards. The owner's purchasing department or legal counsel can usually assist in this determination. Experienced JOC consultants, recommended by professional JOC organizations such as the Center for Job Order Contracting Excellence (CJE), or by consulting services such as those provided by RSMeans, can assist the owner in identifying applicable governing laws, as well as guiding the implementation of a successful JOC program.

Once JOC's applicability has been determined, the planning process can proceed for implementation. The primary steps are as follows:

- Establish a strategic plan that focuses on obtaining "buy-in" to the method from in-house key participants, stakeholders, and interested parties.
- Evaluate internal processes and procedures, as well as staff capabilities.
- Determine which interface type and methods the owner will use to manage the contract.
- Provide training to in-house key participants as needed to develop, award, and execute the contract.
- Determine the availability of contractors who have the resources and capability to satisfactorily meet contract requirements and owner expectations.

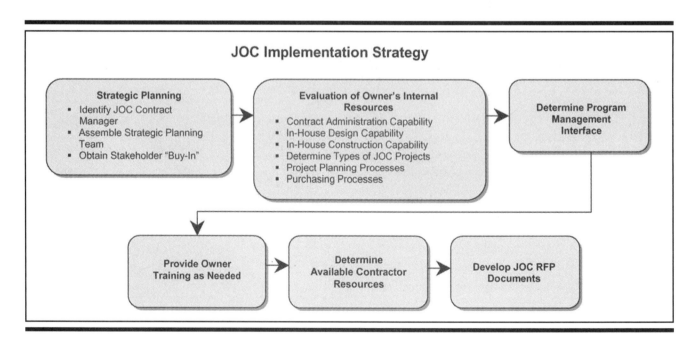

**Figure 3.1** shows one strategy for implementing JOC, including preparation for contract execution as a precursor to the JOC RFP process.

## Strategic Planning

The first step for the owner is to designate one person to lead a strategic planning process that defines the organization's goals, establishes objectives to reach those goals, and identifies actions to achieve the objectives. The owner should develop a plan appropriate to the anticipated volume of work to be awarded. At a minimum, key participants in the process should be trained in basic JOC principles and contract management procedures. For example:

**Goal:** The facilities owner provides construction project requestors with excellent customer service.

**Objective: Implement and execute a successful JOC contract.**

*Action:* Evaluate internal resources.

*Action:* Assemble an implementation team, and determine who will manage the contract.

*Action:* Establish a reasonable procurement time line.

*Action:* Determine available contractor resources.

*Action:* Prepare a legally sufficient JOC contract that reflects the owner's requirements.

*Action:* Solicit Requests for Proposals (RFPs).

*Action:* Evaluate RFPs and award contract.

**Objective: Obtain "buy-in" from key stakeholders prior to implementation.**

*Action:* Identify key participants, stakeholders, and interested parties.

*Action:* Market the main aspects of JOC principles to key stakeholders by explaining how the method will benefit the facility and how this contracting "tool" can be a useful project delivery option.

*Action:* Determine training needs of key participants and provide training.

*Action:* Involve stakeholders in proposal review and the contract award process.

**Objective: Reduce backlog of requests for construction services.**

*Action:* Identify anticipated types of projects to be "JOC'd."

*Action:* Identify a standard specifications manual and list site-specific, common materials and devices. (This will assist the contractor and can reduce JO approval time lines.)

*Action:* Identify and plan for the first wave of projects and sequence volume.

*Action:* Establish processes to accommodate issuance of blanket POs (or similar means used for expediting procurement).

*Action:* Develop an efficient work authorization process.

*Action:* Partner with the contractor to resolve issues and enhance processes.

## Program Management Interface

In today's business climate, computer software applications are essential. JOC requires a designated UPB to enable owner/contractor interface of a JO price proposal. The UPB, with associated software application(s), is used by the JOC contractor and owner to develop and review JO price proposals prior to approval. Other software applications can be used for exchanging various data and communicating with key participants during contract execution.

These basic tools—designated UPB(s) and computer hardware and software applications—are essential for effective program management. When determining requirements for each of these tools, the owner should take into consideration anticipated types of projects to be delivered, annual volume of work, and the service area.

### Unit-Price Book

The ability to accurately estimate the cost of a proposed project is of monumental importance when using JOC. Since the key component for compensation in a JOC contract relies on a designated unit-price cost estimating book, the criteria for selection of the UPB should be nothing less than the most accurate and dependable current resource data available. *(See Chapter 6 for a more extensive discussion of JOC estimating.)*

The unequivocal choice among industry professionals for cost estimating data is published by RSMeans. The basic UPB used by those engaged in JOC for facilities minor construction, repair and remodeling, alterations, upgrades, and so forth, is the most current edition of RSMeans *Facilities Construction Cost Data. (See the excerpts from* Facilities Construction Cost Data *in the Appendix.)* This UPB contains cost data divided into divisions according to the MasterFormat system of classification and numbering as developed by the Construction Specifications Institute (CSI) and Construction Specifications Canada (CSC).

In some cases, multiple unit-cost databases are integrated, depending on the owner's requirements and the types of projects or tasks that are expected to be delivered through the contract. JOC contracts may include additional UPBs, such as the following RSMeans cost data publications and/or electronic databases: *Building Construction, Light Commercial, Repair & Remodeling, Facilities Maintenance & Repair, Mechanical, Electrical, Plumbing, Heavy Construction, Interior,* and *Site Work & Landscaping.* These resources greatly expand the availability of line items, providing increased versatility for matching a line item to a task as needed during JO price proposal development.

However, for most facility JOC applications, *Facilities Construction Cost Data* is best suited as the primary or base unit-pricing reference for the method. If multiple UPBs are designated as allowed, they should be limited in number and directly geared to the owner's facility project applications. If not, there may be confusion as to which line item was derived from which UPB, making price proposal review more complicated than necessary, and possibly inhibiting the approval process.

If an owner has unusual requirements, such as critical "clean rooms," medical research facilities, or other work that is not adequately covered in an existing unit-price book, supplemental line items to cover a facility's specific needs may be developed in-house or provided impartially by others. In these cases, a process must be established to ensure these line items are updated at the same time that the base unit-price book is updated. Sometimes a separate coefficient is bid for these supplemental items to ensure fairness.

## Computer Hardware

Computer hardware for both the owner and contractor should be capable of handling all required software applications. In most cases, the latest versions are not needed, and personal computers with software designated in contract provisions are sufficient. However, for large entities with multiple sites, a server with server-based applications may be appropriate. The owner usually absorbs the cost of the server (if required) and any personal computer hardware or software used for contract administration purposes, including maintenance and upgrades. The contractor is typically responsible for purchasing compatible personal computer hardware and software.

## Computer Software

Software compatibility among JOC program participants is essential for program management interface. Applications do not need to be complex, and those that provide "real-time" information, in particular, can expedite pre-construction activities and enhance contract execution. Requirements for all software should be clearly identified in the contract provisions.

Computer estimating software using RSMeans cost databases is often standard with JOC contracts. For JOC applications, the most current CD-ROM edition of *Means CostWorks®* and *Means CostWorks Estimator™* should be considered. RSMeans also provides quarterly updates available for download, allowing users to stay abreast of current fluctuations in pricing. Means' newest product, *Means JOCWorks™*, is customized for JOC projects and combines estimating and project management features. In addition to importing/exporting estimates in *Microsoft Excel™*, *Microsoft Word™*, and *Adobe™ PDF*, *Means JOCWorks™* supports multiple and dual award contracts and pricing guides and organizes projects by contract, performance period, contractor, location, estimator, or customer. *(See the Resources section at the back of this book for more information.)*

*Microsoft Project©* is also easy to use and is an excellent choice for tracking and scheduling projects, as well as meeting most contract reporting requirements.

Proper training of all software users—especially those who will be engaged in cost estimating—is imperative. The contractor may be required to provide start-up training of the owner's key participants in certain types of software applications that are critical to the program management interface, such as cost estimating. In this case, the number of owner's employees to be trained, as well as either the description of appropriate courses or the minimum number of owner-approved training hours, should be clearly identified in the contract provisions. Other software programs using Means cost data that have been used for JOC are: Pulsar from Estimating Systems, Inc., WinEst® from WinEstimator, Inc., Success from US Cost, Inc., and e4Clicks from 4Clicks-Solutions, LLC.

## Review of Owner Resources

Before implementing a JOC program, the owner should evaluate existing processes and staff abilities to determine whether modifications or training may be needed. Consideration should be given to the following:

- Project planning
- Existing in-house contract management and construction capabilities
- Purchasing
- Anticipated annual value and types of JOC projects
- Available space identified for JOC contractor operations and/or storage (if necessary)

### Planning Projects

One attractive feature of JOC that should be emphasized when promoting it to upper-level administrators is its capability to carry out small projects quickly. A large facility or agency may have a backlog of deferred maintenance projects or requests for the types of projects that JOC is designed to deliver. The owner may or may not have a policy in place to identify and plan for requests for minor construction, renovations, or modifications.

Depending on the volume of requests, it may be difficult for the owner's construction contract manager to effectively schedule projects without some sort of structured planning, authorization, and prioritization process. Consequently, many public and private entities have internal procedures for approval and prioritization of projects before they are submitted for scheduling. This process usually is initiated by a call for project requests to be submitted to facilities planning administrators from requestors through their respective chains of command, during pre-established time frames, to allow sufficient lead time to properly plan for the requested work.

Projects that are smaller in scale and scope that require minimal design effort, such as nonstructural building interior renovations, alterations, and upgrades; building exterior rehabilitation; and site work projects, such as parking lots and sidewalks, are good candidates for JOC. These may have less formal requirements for request and approval procedures than larger projects of greater value or complexity, with submittal of requests closer to the desired start date. However, project requestors (especially for renovations and alterations) may need assistance in developing budgets and estimating the time needed for design and construction. In an ideal contractor-owner relationship, the contractor can be very helpful in this process.

Once planned and prioritized, projects are submitted to the facility's construction department for scheduling through the owner's construction contract manager. The contract manager should also allow sufficient time for emergency and unplanned projects—usually 25% of the anticipated annual volume. The owner's contract manager works with the contractor to schedule pre-approved JOC projects for times that have the least impact on operations. This also helps ensure that projects with critical performance dates are started and completed within desired time frames. The JOC contractor can plan in advance to obtain sufficient internal staffing and availability of subcontracted resources to handle the workload. The result is better, more efficient customer service to the requestor.

## In-House Resources

JOC is not meant to replace an owner's in-house construction or maintenance workforce. There will be times when the volume of work might exceed the capabilities of the in-house workforce, or when it is more efficient to direct in-house staff to emergencies or maintenance needs. JOC can "fill the gap" between projects that in-house construction workforces can readily accomplish and those that are larger in value, scale, and complexity (and best matched to other delivery methods). JOC can also be applied as needed to reduce backlogged maintenance requests and handle surges in maintenance, as contract requirements allow.

When projects are budget- or time-sensitive, JOC allows the flexibility to integrate the contractor's resources with the owner's in-house resources, as needed. A facility owner's in-house construction workforce should feel comfortable with the implementation of a JOC program. Facility administrators and contract managers can help ensure this through implementation strategies that are sensitive to in-house workforce needs, and obtain buy-in from these stakeholders through active participation in the method. Some of the best JOC contract administrators and inspectors have had previous in-house construction and/or maintenance experience. In addition, in-house construction and maintenance staff can provide valuable input as supplemental site visit attendees during project scoping and as reviewers of project plans and written scopes of work.

## Procurement Processes

For a public sector JOC program to achieve its potential for quicker project response times, the owner's procurement process for individual JOs may need to be assessed. In the public sector, issuing a PO to *encumber*, or commit, funds for a construction contract usually requires approval by owner's representatives (sometimes upper-level administrators) before processing by the owner's purchasing department. For example, an individual project may involve approval from city councils for municipalities or from school boards for educational facilities.

Depending on the procurement process, this final approval can take anywhere from one to several weeks—and sometimes longer. This is counterproductive to quick project starts and can inhibit good customer service when quick starts are desired or required.

With JOC, the awarded contract is already in place, and therefore already authorized for execution. Two signature levels above the owner's representative directly involved with the JO pricing review should sufficiently comply with any internal checks and balances typically required for public sector procurement. Often, these signatures can readily be obtained after established review procedures from administrators within the owner's facilities department. This enables the owner to work with the JOC contractor to expedite project starts.

With a PO in place, work usually can commence when the JO is authorized. Sometimes, a *blanket* or *framework* PO to the JOC contractor can be utilized in order to reduce the owner's internal procurement cycle times. The value of each job order is subtracted from the balance of the blanket PO. Unused balances revert back to the funding account at the end of the fiscal year or at the end of the contract term. The amount of the blanket PO will vary with the anticipated annual volume of JOC projects and available funds. It should, however, be at least one-third to one-half of the annual anticipated value of work to sufficiently realize the benefit of reduced procurement time lines.

The PO can be increased incrementally as needed. Amounts can be determined by projecting the cumulative value of anticipated projects within the remaining fiscal year or until the contract term ends. When funding for projects will come from a number of different owner accounts, a *clearing* or *revolving account* (from which blanket POs can be issued) should be considered. These accounts are continually reimbursed by other accounts at job order completion. (Owner requirements associated with purchasing and accounting processes may vary.)

Eliminating the procedure for developing separate contractual documents for each project is an obvious benefit of JOC, since this process, combined with contract solicitation and procurement, requires significant time, expense, and effort. Once a JOC contract is in place, not having to issue individual POs for each JO streamlines approval processes and reduces procurement cycle times even further.

## Owner Training

JOC's unique processes (when compared with traditional project delivery methods) require training for key participants, as well as sufficient knowledge about the method for those who are indirectly affected by its implementation. The owner's training efforts for successful implementation of a JOC program should include, but not be limited to, the following:

- Designating one person as a "JOC Contract Manager" to champion the program's implementation and provide focused leadership. This person should have experience in construction contract management and proven supervisory skills and should possess the authority to make day-to-day program decisions.
- Distributing information about JOC methodology and attributes to all stakeholders, thereby promoting buy-in. The focus should be on basic JOC procedures and how JOC will fit into the facility owner's toolbox of project delivery methods.
- Identifying JOC's primary attributes, such as accelerated pre-construction time lines and stable, predictable pricing structures.
- Training staff directly engaged in JOC contract administration. Staff should understand JOC principles and the aspects of partnering, plan reading and material takeoff, unit-cost estimating, basic principles of negotiation, and all provisions of a JOC contract.

JOC seminars and conferences from credible sources can also be a good source of training for owners who are considering program implementation, as well as contractors who may be engaging in the contract. Training events also benefit those who already have JOC programs in place and want to make contacts in the industry, share experiences by networking, or stay abreast of current JOC issues.

## Contractor's Review of Owner's Resources

Establishing JOC as a preferred delivery method (for projects well-matched to it) is not "business as usual" for owners or contractors. As soon as it is apparent that an owner is seriously considering JOC implementation (or at the time the RFP is posted), contractors should begin to review the owner's resources to more clearly understand their operational processes, contract management abilities, and the volume and types of projects expected to be administered through the contract. These issues will affect coefficient development and contractor requirements—including response times.

Contractors should address the following issues:

1. *Has the owner identified a single point of responsibility for the program?* If not, there may be inconsistency in contract administration, as well as delays in decision making.

2. *Does the owner have internal design capability?* If not, and depending on contract requirements, the contractor may need to provide design services. This includes code review drawings and/or A/E services provided by or subcontracted through the contractor.

3. *In addition to the types of projects the owner needs accomplished and their anticipated volume, to what extent can the owner plan future work?* The owner must be able to provide sufficient notice to the contractor for pending work.

4. *Are there certain times of the year that the owner will need additional contractor resources to handle peak project loads? Are there times when little or no work is scheduled by the owner?* Answers to these questions can assist in determining the potential for fluctuating required resources to match the owner's project volume.

5. *What are the owner's needs in regard to emergency work?* This may impact the contractor's in-house workforce or subcontractor management.

6. *What is the anticipated workload to be performed during non-standard hours?* This may impact the contractor's ability to provide flexibility in performance times.

7. *To what extent can the owner provide timely record drawings?* This will impact JO proposal submittal response times.

8. *What are the typical subsurface conditions of the owner's facilities?* This will give the contractor an idea about what kind of equipment will be required for site work projects—especially excavation and ditching.

9. *Does the owner have documentation of existing hazardous building materials?* This can impact project costs, completion dates, and regulatory compliance.

10. *What is the owner's requirement for work permitting, and who will be responsible for obtaining permits?* This affects pre-construction time lines.

11. *What is the owner's service area?* Some owners may have centralized, compact facilities with staging and access limitations. Others may have numerous facilities spread over large metropolitan areas, or the contract might include riders with agreements between other entities, allowing access to the contract. In the latter two cases, workers may spend a significant amount of travel time to get to work sites. Some owners may have facilities in rural areas with little or no local resources available. Logistical costs may need to be included in the contractor's coefficients, depending on the contract's pricing structure requirements.

12. *If riders are included in the contract, what is the total value of anticipated work to be performed?* This information can impact coefficient development.

13. *Will the owner require full-time supervision at each project site?* This can affect field supervision costs and may need to be included in coefficients.

14. *How does the owner budget work, and what are the processes of work authorization?* These issues impact the owner's ability to provide a consistent volume of work, as well as the time it takes to review and approve JO proposals.

15. *Are there any politics involved in the JO award?* Pre-construction time lines may be affected if the owner's JO must be approved by various boards, committees, or councils. The time it can take for authorization to proceed with work as scheduled may be surprising to contractors who have not dealt with public or government entities. Also, if the final decision-making process for day-to-day operations is at a high administrative level, work in progress may be delayed if there are changes in the scope of work or job conditions.

16. *Are there other contracts that may be used for similar type work?* Contracts that may impact the amount of available work include single-trade ID/IQ contracts, time and material contracts, small value service provider purchase via single source procurement contracts, and other contracts alternative to JOC.

17. *What is required for invoicing, and how long does it take to receive payments?* This can impact the contractor's available cash reserves and bonding capabilities.

18. *Is the owner a private corporation, privately owned entity, or public-sector entity?* Public-sector entities usually have more stringent reporting requirements, as well as more formal processes for work authorization and payment issuance.

Information about the owner's operational processes, staffing resources, and facility characteristics can assist the contractor in aligning internal staffing and subcontracted services to match contract requirements. This information can also be a major component in coefficient development. Many of these issues can be addressed during the owner's pre-proposal conference.

## Owner's Review of Contractor's Resources

Just as contractors considering a response to a JOC RFP should assess the owner's resources, owners will need to conduct their own review of JOC contractors. Owners should evaluate the abilities of local contractors (versus larger, specialized JOC contractors) to successfully engage in the method and manage the volume of work.

The formal process of contractor evaluation is conducted during the RFP selection process. However, this issue might be preliminarily addressed even before RFP development to ensure contract provisions are tailored to meet specific owner requirements. Matching the owner's requirements with the capabilities of the contractor is essential to the success of a JOC program. Consideration should be given to the following issues:

1. *Does the RFP provide the opportunity for local contractors to submit competitive proposals?* Some owners feel it is in their best interest to build JOC relationships with local contractors, subcontractors, and vendors with whom they are already familiar from other projects and delivery methods. Local contractors must be willing to learn the principles of JOC.

   On the other hand, larger, specialized JOC contractors are already well-trained, typically have greater bonding capacity and financial resources, and can usually offer a much wider range of services to the owner, such as code review drawings and/or A/E services. Usually, they also have a greater capacity for handling large volumes of work concurrently, and sometimes they can offer a more favorable price proposal than smaller local contractors. Price proposals offered by large contractors can reflect a smaller percentage of profit margins as compared to smaller contractors.

   According to RSMeans, profit margins can range from 35% for a small contractor with annual sales less than $500,000 to 5% for a large contractor with annual sales in excess of $100 million. An owner may need to balance the issue of large contractor capability and competitive pricing against retaining direct contractual relationships with smaller local contractors of proven performance. However, by requiring plans to maximize the use of local small businesses to include historically underutilized businesses (HUBs), the owner may be able to obtain the advantages of both a large JOC contractor and local business utilization. The evaluation criteria must reflect and score this requirement to ensure this need is met. (Owner requirements for HUB participation may vary.)

2. *Does the contractor directly employ a labor and/or trades workforce?* JOC contract provisions may require a certain percentage of contractor-employed workforces (25% or so)—or at least one multi-disciplined crew of three to five tradespersons. This is because some owners feel that the contractor's direct control of workforces can enable more timely responses for projects requiring quick mobilization or emergency requests.

Some owners may also feel that direct control of workforces gives the contractor a more vested interest in the program's success. On the other hand, requiring the contractor to have a certain percentage of in-house workforces may not be important to other owners.

## Contractor's Review of Own Resources

Contractors must be familiar with qualified, dependable, local trade subcontractors and vendors—and determine whether they can sufficiently assist in providing JOC services to the owner. Larger, specialized JOC contractors often seek out area subcontractors, or even other small general contractors, who have a reputable history and are capable of performing quality work when needed. If labor resources are not available locally, they will have to be brought in from other areas—usually at a greater expense. Contractors should review potential candidates who can be mentored as JOC participants.

Following are additional actions the contractor is advised to take:

1. *Examine the knowledge, skills, and abilities of key staff members who will participate in the JOC program.* Sometimes contractors discover that staff members proficient in other types of delivery methods may not be well-matched for the partnering aspect of JOC. Consideration should be given to staff with a "team player" attitude, as well as a desire to provide quality customer service. JOC contractors have seen contracts fail when experienced project management staff lack the interpersonal skills needed to effectively service a JOC contract.

2. *Evaluate the financial status and bonding capability prior to RFP response.* This ensures these capabilities are sufficient to execute the contract and meet designated bonding requirements. The contractor should also be aware of costs associated with mobilization and start-up—and have adequate financial resources in place to provide services to the owner as required. For public-sector work, it may take as long as 90–120 days from initial JOC program start-up to receive payment for the first submitted pay application. In the interim, the contractor may have substantial financial resources committed.

## Contractor Training

The contractor's training efforts for a successful JOC engagement should include, but not be limited to, the following:

1. Designating one person within the organization to learn and practice JOC principles and serve as a single point of responsibility to the owner

2. Developing a coefficient that allows reasonable compensation for services rendered

3. Training all key staff members and subcontractors in applicable provisions and referenced documents of the JOC contract

4. Training respective management and technical staff in:
   - Developing an accurate scope of work (critical)
   - Plan reading and material takeoff
   - CADD drafting
   - Unit-price estimating
   - Scheduling

- Developing required status reports
- Engaging in effective contractual agreements with subcontractors
- Project management
- Construction processes
- Negotiation processes
- Quality control
- Job site safety
- Developing and compiling required close-out documents

## Conclusion

Making sure that key participants and support staff who will be responsible for a facility's construction efforts fully understand and support the principles of JOC is essential to successful implementation. This, in conjunction with other implementation actions identified in this chapter, are part of a thorough JOC program strategic planning process. They will help ensure the method gets off to a good start at the owner's facility.

Identifying the best available program management interface components to meet the needs of the owner and the contractor—especially the UPB and associated cost database software applications—will eliminate (or greatly reduce) the likelihood of JO price proposal negotiation problems. Designated cost data must be reliable, consistent, impartial, and easy to understand and use. Problems with JO price proposal negotiations due to poor UPB choice or cost data misapplication can sour relationships by dissolving trust. These problems can prevent the partnering process from reaching its full potential or may even cause the relationship to fail.

For other delivery methods, evaluating resources and conducting training may not be necessary—but JOC is not just another delivery method. Most likely, the owner's contract manager and administrators are already well-trained and efficient in their duties and have some knowledge of the principles of JOC, which will help make program implementation flow smoothly.

Joint training sessions in estimating, negotiating, and technical skills for both the owner's and the contractor's key participants should be considered after contract award. This can be beneficial not only in enhancing their abilities in these areas, but in helping promote an atmosphere conducive to partnering. If desired by the owner, an experienced JOC consultant will train the owner's key staff members in implementation and execution procedures.

# Chapter 4

# Developing a JOC RFP

JOC contract documents are similar in structure to those of most other delivery methods and feature many of the same conditions, exhibits, and attachments. However, several sections are distinctly different, most notably the proposal documents to be developed and submitted by the respondent (contractor) for evaluation, and the General Requirements for JOC, which can substitute for the "Project Description" section of a typical project-specific RFP.

The respondent's management/technical proposal (sometimes referred to as the *performance proposal*) is typically more extensive than those found in other RFPs, as more emphasis is placed on it during the evaluation process. JOC RFPs may also require respondents to submit a sample JO proposal—with code review drawings, detailed scope of work, and price proposal—as a sample of their abilities. Respondents should be allowed ample time to prepare proposal documents in response to a JOC RFP, typically a minimum of three weeks.

## Research vs. Reinvention

Templates for JOC RFPs can be obtained from either model contracts from sources such as CJE or from entities already successfully engaged in JOC. *(See the "Resources" section at the back of this book.)* Owners usually customize JOC contract provisions to suit their individual requirements. To get a feel for basic JOC contracts, it is a good idea to research existing JOC RFP documents with similar requirements and consider the content as it might apply to the owner's facility.

Sufficient time should be allotted by the owner's contract management staff to review sample contract documents and carefully consider any deletions, modifications, or additions. Once drafted, the contract document's "boilerplate" and other related sections may reveal how the provisions correlate to those specific to the JOC General Requirements section. Sometimes there are conflicts of provisions within related documents. In these cases, a statement may be included identifying the hierarchy of applicability among similar contract provisions. Owners who decide to utilize JOC should communicate with other owners already successfully engaged in JOC at other facilities during contract development. The insights they gain may be helpful in developing their own JOC RFPs.

JOC contract documents should be reviewed by a representative of the owner's purchasing department, as well as by an independent consultant with JOC expertise. The consultant should have experience with applicable jurisdictional requirements regarding JOC implementation. Once the owner's contract provisions are identified and drafted, the documents should be reviewed by a legal professional. Attorneys with a background in construction law can help ensure compliance with any governing statutes or regulations, as well as current case law decisions. Legal service providers can also assist in identifying potential risks to the owner—and make recommendations to transfer, mitigate, or share those risks.

## RFP Solicitation

Solicitations for RFPs must be advertised or posted in accordance with applicable jurisdictional purchasing requirements. Some owners may target builders' exchanges, plan rooms, or specific contractors' organizations. Only complete RFP packages should be issued. When issuing RFP documents in paper form, a refundable deposit is typically required from the respondent to help cover reproduction costs. CD-ROMs in "read only" format minimize the use of paper products, as well as the expense and time spent making and delivering copies. Internet postings with links to downloadable RFPs are also becoming common in the industry.

It should be noted that the selection method used to award a JOC contract may involve an RFP process, an RFQ, or an RFQ followed by an RFP. The following are examples of methods that may be used to award a JOC contract. Any method used in the selection process should be sufficiently reviewed for compliance with the owner's purchasing requirements.

- *RFP process:* evaluates all qualified proposals. Selection is based on the highest-ranked respondent's management/technical proposal, price proposal, and sample JO proposal (if required).
- *One-step RFQ process:* identifies a short-list of respondents based on their qualifications. Submittals include a management/ technical proposal and sample JO proposal (if required). Selection is based on negotiations with the highest-ranked respondent. If an agreement cannot be reached, the next most qualified respondent may negotiate with the owner. The price proposal is not a factor for selection in this process.

- *Two-step RFQ process:* identifies a short-list of respondents to the RFQ who are invited to submit a response to the RFP. The price proposal is a factor for selection.

## Pre-Proposal Conference

Contractors who would like to submit a proposal in response to a JOC solicitation should attend a pre-proposal conference. For public-sector entities, if allowed by applicable jurisdictional authorities, attendance should be mandatory. The conference is typically conducted by authorized owner's representatives, and the JOC contract manager should be present to answer any questions.

The following issues should be covered:

- The basic contents of the proposal documents to be submitted, with emphasis on contractor qualifications and performance abilities
- Any required proposal guarantee or bid bond, or payment and/or performance bonds
- All required forms to be completed and included in the proposal with emphasis on price proposal requirements
- Any other subject deemed relevant by the owner or the respondents, such as owner facility characteristics, operational processes, and JOCable project data

If respondents are new to JOC, the owner should be prepared to conduct a training session covering basic principles of the method. Owners implementing a JOC program for the first time might bring in an experienced JOC consultant to assist with the conference. Handouts that list and briefly identify the types of potential JOC projects (for both standard and non-standard hours, based on actual historical data) can be valuable to respondents. (Sometimes this information is attached as a reference document in the RFP.)

RFP respondents will need to know the owner's projection for the annual value of JOC contract work—and whether there will be multiple contract awards issued for it. This information will help respondents proportionally calculate their own anticipated volume of work, a significant factor in their coefficient development.

The pre-proposal conference should also address the owner's processes of requesting job order proposals from the contractor, review and approval procedures for job order authorizations by the owner, and invoicing and payment procedures.

## RFP Overview

Properly prepared JOC RFPs outline the following for respondents:

- When and where their proposals will be received
- Contract terms and conditions with attachments for reference, including JOC General Requirements
- Proposal requirements and instructions on how to develop proposal documents

- Evaluation criteria used for award
- Exhibit documents to be completed and submitted

Figure 4.1 identifies general JOC RFP components and related items. The "Proposal Documents to Develop and Submit" category includes components of the management/technical proposal. For a JOC RFP, these documents, along with the respondent's price proposal, make up the primary criteria for evaluation purposes. General Requirements of the JOC, under "Contract Terms and Conditions," will be addressed separately with sections specific to JOC addressed in detail. (See Chapter 5 for a more extensive discussion of JOC General Requirements.)

## JOC RFP Components

JOC RFP categories and characteristic components are summarized in the following list:

1. **Request for Proposals:** The written advertisement that announces the request for contractors to provide sealed proposals for JOC services to the owner for consideration. It identifies the owner and includes a brief description of JOC—including identifying JOC as an *indefinite delivery/indefinite quantity*-type contract. Also included is information on how to obtain proposal documents, dates and times for the pre-proposal conference, and the proposal submittal deadline.

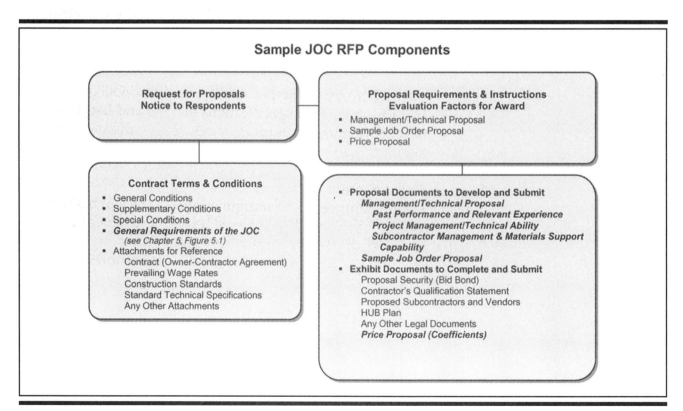

**Figure 4.1** illustrates sample JOC RFP components. Actual owner requirements may vary.

2. **Notice to respondents:** Expands the RFP and may include a description of the owner's facilities and contact information for inquiries and interpretations. A brief description of contractor eligibility requirements, criteria for award, and the award process are provided. Other general information, such as contract terms, minimum guaranteed annual volume, bonding requirements, and commitment or obligation statements, is summarized.

3. **Proposal requirements and instructions, and evaluation factors for award:** A section of the RFP that identifies the documents to be developed and submitted by the respondent. Detailed instructions are provided for how to prepare these documents. Respondents submit a management/technical proposal consisting of information regarding their relevant experience, project management and technical abilities, and subcontractor management and materials support capability. The management/technical proposal requirements identify specific contractor information to be submitted:

    - Project history and references
    - Relevant experience with JOC and/or the types of projects that the method is designed to deliver
    - Current qualifications of key management and field staff
    - Financial capability
    - Proposed plan for managing work
    - Ability to provide subcontracted resources and materials

    A sample request for a job order proposal may be included to evaluate respondents' abilities to formulate written responses to a request in accordance with the owner's contract requirements. The price proposal for the RFP is referenced as a form to be submitted for evaluation. All forms to be used by the owner in the RFP evaluation process are attached for reference.

4. **Contract terms and conditions:** The contract's General Conditions and Supplementary Conditions. Special Conditions of the contract, which outline provisions or articles specific to JOC General Requirements and may provide clarity to General and Supplementary Conditions, are also included. General Requirements of the JOC contract can reside in this section, as shown in Figure 4.1. These contract provisions relate to all Division 1 General Requirements. Any other reference documents linked to contract terms and conditions may be attached, such as the contract, or *Owner-Contractor Agreement*; prevailing wage rates; the owner's facility-specific Construction Standards and/or Standard Technical Specifications (reference unless short); a sample JO authorization form; and a change order form. A list of projects (with brief description and values completed over the preceding 12-month period) is also useful for respondents.

5. **Exhibit documents to complete and submit:** All forms to be completed by the respondent and included in the proposal, including the proposal security (bid bond), contractor's qualification statement, proposed subcontractors and vendors form, HUB plan (if required), price proposal form, and any other documents required by the owner to be completed and submitted.

## Management/Technical Proposal

The management/technical proposal consists of three sections of information about the respondent, to be developed and submitted in response to the RFP:

- Past performance and relevant experience
- Project management and technical ability
- Subcontractor management and materials support capability

Respondents will also need to submit a sample JO proposal (if specified) and a price proposal. Owner requirements for management/technical proposals may vary.

### Past Performance and Relevant Experience

This section details the respondent's past performance—an important factor in determining capability and reputation. It should include the following:

- The number of years in business
- Contact information for past clients, major subcontractors, and vendors
- A list of projects completed over a designated period of time, with emphasis on those that have similar value, scope, and scale to the jobs anticipated to be awarded by the owner
- Demonstrated ability to perform work on multiple projects simultaneously while engaging in contractual relationships with multiple subcontractors

Other information about the respondent's past performance may include, but is not limited to, the following:

- Documented certification of good standing with any jurisdictional government or regulatory authorities that may track the ineligibility of contractors
- Response to requests for project proposals and ability to meet desired project start and completion time lines
- Provision of technical assistance, constructability methodologies, and other cost-saving measures, such as value engineering, to help maintain individual project budgets

- Service as a team member and/or experience partnering with key project participants
- Submission of contractor-generated change orders and claims. (A history of many claims may raise a "red flag.")
- Record of meeting goals or requirements associated with providing outreach, mentoring, and subcontracting opportunities to businesses considered historically underutilized or disadvantaged by social, economic, or physical factors
- Safety record

Owners may choose to require that the past performance and relevant experience portions be submitted before the rest of the management/technical proposal so that the evaluation staff may start checking references, performance information, and experience while the respondents are completing the rest of the proposal. This will reduce the total time required to evaluate the proposal.

**Project Management & Technical Ability**

This section of the management/technical proposal is always an area of primary consideration in JOC proposal evaluation processes. The respondent must consistently provide management and technical staffing resources that will meet the owner's requirements. Substitutions of any key project management or technical staff identified in the proposal after contract award without compliance to owner pre-approval requirements may constitute a basis for dispute.

The respondent should include the following:

- **Staff and company information:** A detailed organizational chart listing all key staff members, with emphasis on primary points of contact for the contract and technical services. If the organization is of a corporate nature, its business profile should be included. Qualifications of individuals who will participate in meeting contract requirements should also be submitted, including their education and relevant training, skill levels, credentials, affiliations, past experience, and levels of authority in the specific roles they will play. Technical support, such as estimators, QC staff, CADD drafters, or professional consultants, should be identified.

  With JOC's pricing structure, the estimator's experience, ability, and competence are of particular importance throughout the term of the contract. In-house design, trades, and labor workforces and their qualifications should be listed, as well as any clerical staff to be assigned to the owner's account.

- **Business profile:** A written account of the respondent's business profile, how and why the respondent's services are well-matched to the owner's requirements, and any conditions that may impede a contractual relationship.
- **Quality control plan:** A plan identifying all aspects of inspection, acceptance, and rejection of work, including an approach for quickly documenting and correcting deficient quality. How and when the contractor communicates quality issues with the owner's contract administrators should be explained. A narrative should also be included to clarify the methodology to be used for site housekeeping and final cleaning.
- **Financial profile:** Respondents must demonstrate financial stability and documented evidence of their ability to obtain the required bonding and insurances. They must also identify cash reserves. A plan explaining how start-up costs will be absorbed prior to payments for work accomplished may be required, as well as a payment plan for subcontractors and material suppliers. Cost control methodologies (to ensure the respondent's internal financial stability over the term of the contract) may also be required.
- **Procedures for project management and technical services:** These should be specific to the owner's requirements and emphasize the respondent's work plan, implementation strategy, and ability to effectively schedule and execute the owner's anticipated concurrent volume of work. Sample schedules and reports (using required software) must be submitted.

Other requirements for demonstrating project management and technical application methodologies may include the following:

- Methods of response to requests for JO proposals to ensure response requirements are met. A contingency plan should be included to accommodate sudden increases in workload and response to emergency requests
- Method for developing code review drawings, scope of work, and JO price proposals, in accordance with the General Requirements of the JOC contract, including software applications
- Procedures for communicating with designers, subcontractors, and owner's representatives and clients
- Health, safety, and first aid procedures including use of personal protection equipment (PPE)
- Environmental control plans for both inside and outside of buildings. These should describe the environmental health and safety control measures that will be taken to protect building occupants. For outside work, the plan should include procedures for work site security, appropriate signage, traffic control, pedestrian protection, and noise and dust control.

**Subcontractor Management & Materials Support**

The contractor must identify a plan to effectively utilize subcontractors and materials suppliers. The plan should describe the selection and award process and explain methods of managing and coordinating the work of subcontractors after award. Emphasis should be placed on how the contractor will secure and retain effective relationships with subcontractors in an environment that often requires quick response times (for both contractor-subcontractor agreements and mobilization).

Materials suppliers should be identified, including those who can pass along materials, product data, and samples to the contractor in a timely manner for subsequent submittal to the owner. Other information to be submitted may include the following:

- A list of proposed subcontractors for each major trade discipline, including a description of the types of services they provide, their experience, and their contact information
- The anticipated percentage of subcontracted services (by trade) in relation to in-house workforces
- A plan describing how reports of subcontractor activity will be communicated to the owner as required
- A plan outlining how the contractor will provide assistance to subcontractors regarding payroll and materials acquisitions, if necessary
- A narrative describing the contractor's method of obtaining professional consulting services, if applicable
- A description of training or mentoring programs provided to subcontractors (including the use of HUB subcontractors and vendors)
- A specific plan for identifying and using HUB subcontractors and meeting the agency's goals (frequently required by government agencies)

## Sample Job Order Proposal

A sample JO proposal can be a submittal requirement to give evaluators an idea of each respondent's technical abilities. A representative project that can be JOC'd is identified. (It can be a project that has already been completed or one that has potential to be authorized for completion during the contract term.) A description of the project, along with associated design documents and specifications, is included in the RFP.

Using owner-furnished design documents and specifications, each respondent must generate a written scope of work and price proposal developed from applicable UPB line items and designated for use in the JOC General Requirements section. The price proposal component of the sample JO proposal should be developed with the respondent's proposed coefficient, and in accordance with the pricing requirements of the JOC General Requirements section of the RFP.

An alternative to furnishing respondents with design documents is conducting a site visit to a representative JOCable project at the owner's facility. This event can be a part of the evaluation process. The site visit can include all pre-proposal conference attendees, or only those on the short-list of potential RFP respondents. The respondents would then develop a code review drawing, a written scope of work, and a price proposal for the representative project. This would be done in accordance with related requirements of the RFP.

Either of these two sample JO proposal submittal exercises can be used as an effective evaluation tool to ascertain the technical abilities of the respondents. This gives the owner a clear idea of how well the respondents can scope a project and develop an accurate JO proposal in a timely manner, and how well they understand the contract requirements.

### *Price Proposal*

The price proposal form is an exhibit document for each respondent to fill out and submit. It contains the coefficients offered in response to the RFP. The number of different coefficients required for proposal may vary depending on the owner's needs. Typically, at least two coefficients are required—one for work to be performed during standard or normal working hours, and one for work during non-standard hours. (The coefficients are in relation to work performed using the assigned wage rate category.)

When work requires either the owner's jurisdictional prevailing wage rates or *Davis-Bacon* rates, two additional coefficients may be required (again determined by standard or non-standard working hours). Other coefficients to be proposed by the contractor, such as those for remotely located facilities, those as "riders" to service other owners' facilities, or those for maintenance work, may be included in the basic requirements of the RFP. They may also be designated to be bid as alternates—with or without evaluation weighting.

Some JOC contracts require a detailed breakdown of proposed coefficients into individual cost data components relating to overhead, markups, and profit margins. This breakdown helps clarify how the coefficients were developed and may assist evaluators in determining the value and completeness of the offer.

## Addenda to the RFP

Addenda may be issued prior to the receipt of proposals. Addenda are typically distributed only to those potential proposal respondents who have requested proposal documents, and/or those who have attended the pre-proposal conference.

# RFP Evaluation & Award

The criteria for JOC proposal evaluation may vary depending on the owner's specific needs and compliance with jurisdictional purchasing requirements. A successful evaluation process is lawful, thorough, fair, and impartial. To this end, criteria for proposal evaluation should be clearly outlined in the RFP. The intent of the evaluation process is to determine which respondent(s) is (are) most capable of offering the best overall value to the owner. The evaluation process should conclude as soon as is practical after the receipt of proposals, usually within ten days.

Evaluation committees often determine the JOC contractor selection and award process. The number of committee members may vary, but typically there are three to seven individuals who rate each qualified proposal via a point score. At least one staff member from the owner's purchasing department should serve as an evaluator, along with other stakeholders in the contract execution. The JOC contract manager should abstain from evaluating to remain impartial. During this process, it is best if evaluators refrain from direct communication with each other about individual proposal content in order to ensure objectivity.

Some JOC RFP evaluation procedures include interviewing the respondents on the short-list. In this case, the contractor's and owner's respective key JOC participants should be in attendance, together with the evaluation committee members.

## Evaluating Management/Technical Proposals

Management/technical proposals should reflect all information required in the preparation instructions. They should address proposal response requirements, organized in a pre-designated tabbed format for ease of reference. Coverage should be clear, concise, in accordance with instructions, and well-organized. Proposals that are too general or contain unnecessary content may be rejected. Proposals may also be rejected if required materials are missing or if they contain inaccurate contact information.

Evaluators often contact the listed references to verify the contractor's quality of work and reliability. To expedite the evaluation process, references can even be checked by the owner's clerical staff using a questionnaire format developed specifically for the task. Copies can then be submitted to evaluators for consideration. Financial references, including bonding capability, will also be verified.

## Sample JO Proposal Evaluation

The sample JO proposal contains the contractor's detailed scope of work and JO price proposal with detailed UPB breakdown. Emphasis should be placed on how accurately the scope of work reflects the representative project drawings and specifications. The JO price proposal should reflect the scope of work in relation to the allowable

UPB line items identified in the designated JO pricing structure—including the respondent's proposed coefficient applied to the sum total of UPB line items. The ability to submit a comprehensive and accurate sample JO proposal, in accordance with the JOC General Requirements section of the RFP, serves as an objective criterion for evaluating the respondent's technical abilities.

## Price Proposal Evaluation

When multiple coefficients are required, a weighted composite can be developed. This will assist in the evaluation process (although award can be based on each coefficient proposed, as well as the composite, whichever is deemed most advantageous to the owner). The sample price proposal in Figure 4.2 shows four coefficients, as required for applicable JO wage rate categories and owner-approved performance times. The coefficients and weightings are presented as examples only and have no bearing on those a respondent might submit in response to a JOC RFP—or weights an owner might assign to any particular wage rate category.

---

### Price Proposal Evaluation Method

| | Contractor's Coefficient Offer* | | Weight** | | Weighted Coefficient |
|---|---|---|---|---|---|
| A. PWR Standard Hours (.8 x .9 = W) | 1.00 | x | .72 | = | .72 |
| B. PWR Non-Standard Hours (.8 x .1) | 1.20 | x | .08 | = | .096 |
| C. Davis-Bacon Standard Hours (.2 x .8) | 1.10 | x | .16 | = | .176 |
| D. Davis-Bacon Non-Standard Hours (.2 x .2) | 1.30 | x | .04 | = | .052 |
| | | | (1.00) | | |
| Total Composite Price Proposal (A + B + C + D) | | | | = | 1.044 |

---

**Figure 4.2** illustrates a composite price proposal (CPP), developed when multiple coefficients are offered by the respondent in accordance with RFP price proposal requirements. The total CPP represents the sum total of weighted coefficients, based on the owner's anticipated usage of each coefficient category, resulting in a single composite coefficient. This coefficient is to be used only in evaluating each respondent's RFP price proposal—not for JO pricing.

*This column reflects the coefficients offered by the respondent, typically carried to no more than four decimal places, as instructed in the RFP.

**This column reflects the weights assigned by the owner. Each row's weight reflects the owner's approximate anticipated annual value of JOCable work under each applicable wage rate category, carried to two decimal places. The sum total of all weightings should equal 1.00. The weightings in this example are for an owner whose annual anticipated value of JOCable projects reflects 80% of the prevailing wage rate (PWR). 90% of the work in the PWR category is anticipated to be performed during standard working hours, and 10% is anticipated to be performed during non-standard hours. The remainder of the anticipated volume (20%) is shown using Davis-Bacon wage rates. 80% of this remaining work is anticipated to be performed during standard working hours and 20% during non-standard hours. When using this evaluation method, or a similar variation, be advised that an unrealistically low coefficient applied to a little-used wage rate category will affect the CPP.

The respondent's CPP is included as an evaluation component within the RFP Evaluation Form. (See the Appendix.)

## RFP Evaluation Summary

Current RFP evaluation trends include the use of an evaluation summary form, developed by the owner, with sequentially categorized sections and subsections that reflect the respondent's abilities and capabilities. The summary is a matrix that aligns with required RFP proposal format and content, as indicated in the instructions to respondents. Individual evaluators' scores are averaged, normalized, and categorically weighted in respect to each management/technical proposal component, the sample JO proposal, the price proposal, and the results of the owner's interview with the respondent's key participants. The evaluation summary is used to compare respondents and could also be used as a tool to monitor the awarded contractor's future performance.

Individual weighting of each evaluation component varies with the owner's requirements. However, since JOC is a requirements-based delivery method, it is not unusual for evaluators to weight the respondent's management/technical proposal and sample JO proposal (and perhaps the interview) more heavily than the respondent's price proposal. As previously noted, some JOC proposal evaluation procedures do not consider pricing as a factor in the award process, as is the case with one-step RFQs.

Once the proposal evaluation process is completed, the respondent whose proposal is determined to be the best overall value—or most advantageous to the owner—is accepted. The contract document, designated as a reference document in the RFP, is signed by the owner and the contractor, thereby establishing the contract award, pending verification of bonding, insurance, and any other award-related documents the owner requires.

Once awarded, the contractor's proposal documents coexist with and link to the contract documents for the duration of the contract term, subject to any changes implemented by way of bilateral agreement. If there are any discrepancies or conflicts between the contractor's proposal documents and the contract requirements, the contract requirements supersede.

The owner should notify all unsuccessful respondents of the award, in writing, within a few days. With the JOC contract in place, the owner is ready to execute the contract.

## Amendments to the JOC Contract

A unique feature with JOC, due in part to the potentially long term of the contract, is that contract provisions can be amended after award through bilateral agreements between the owner and the contractor. This enables the contract to undergo minor changes in order to enhance processes. For example, additional services can be added to the contract, such as building and/or grounds maintenance. Any proposed bilateral amendments, deletions, or additions to an active contract should be reviewed by a legal professional with knowledge of jurisdictional purchasing requirements prior to adoption.

## Conclusion

JOC RFP development requires research in order to align contract provisions with owner requirements. Development of this document lays the groundwork for a successful solicitation and award process. Submitted proposals should clearly reflect the respondents' records of performance and current capabilities. A thorough and impartial evaluation process in accordance with governing procurement processes will determine the best overall value to the facility owner for JOC services—achieving program implementation objectives.

Once the contract is in place, the owner and the contractor have a joint and continuous responsibility to assist each other—to *partner*—for the term of the contract. This partnership is a key ingredient in the facility owner's ability to reach the goal of providing construction project requestors with excellent customer service.

# Chapter 5

## General Requirements of the Job Order Contract

JOC impacts all processes and procedures prior to the construction phase, beginning with the owner's strategy to implement the contract. Once construction begins, there is little difference between JOC and other delivery methods—with the exception of pre-established methods for pricing changes to JOs and the beneficial results of partnering. This chapter focuses on those aspects of JOC that affect contractor requirements prior to the construction phase (such as design, pricing, and JO authorization processes), but it also addresses project administration, quality, and other general requirements associated with the construction phase. *(Note that requirements may vary by owner. Those shown in this book's examples would be additional to any documented commitments made by the contractor in response to the RFP.)*

The General Requirements (or Division 1) of the contract are JOC-specific and provide clarity to all other contract requirements. If there is any conflict between the General Requirements and other provisions, articles, or requirements of the contract, Division 1 should supersede. The General Requirements amend or replace other contract definitions that refer to "Project Description" or "Plans and Specifications." *(Note that any references to contracts in this chapter are intended to reference JOC contracts, unless otherwise indicated.)*

This chapter will address the subdivisions and sections of JOC General Requirements shown in Figure 5.1. Those that are unique to JOC will be covered in greater detail unless they have already been examined in previous sections of this book.

## General Requirements of the JOC

- Summary of Work
  - Contract Terms & Dollar Amounts
  - Definitions
  - Program Management Interface
  - Professional Consultant
    - Architectural & Engineering Services
- Price & Payment Procedures
  - Job Order Pricing
  - Change Order Pricing
  - Emergency Work Pricing
  - Overtime
  - Contingencies
  - Coefficient Adjustments
  - Invoicing & Payment
- Administrative Requirements
  - Project Management/Coordination
    - Job Order Authorization
    - Contractor Requirements
    - Meetings
    - Submittals
    - Shop Drawings
    - Cutting & Patching
    - Regulatory Requirements
      - Codes & Permits
      - Code Review Drawings
    - Special Project Procedures
    - Field Personnel

- Construction Progress Documentation
  - Status Reports
  - Scheduling
- Quality Requirements
  - Quality Control
  - Inspections
- Temporary Facilities & Controls
  - Construction Start-up Period
    - Temporary Utilities
  - Construction Facilities
    - Temporary Construction
    - Construction Aids
    - Vehicular Access & Parking
    - Barriers & Enclosures
    - Project Signs
    - Equipment Rental
- Execution Requirements
  - Cleaning & Rubbish Handling
  - Job Order Closeout
- Facilities Operations
  - Operations & Maintenance
  - Facilities Maintenance
  - Moving Equipment

**Figure 5.1** shows component General Requirements for the contractor to perform in accordance with a JOC contract. The subdivisions are presented as an example only. Individual owners should develop their own contract provisions.

## Summary of Work

This subdivision of the General Requirements addresses contract terms and amounts, definitions, program management interface, and professional consultants (if applicable). It defines JOC requirements and specifies the types of projects to which the owner intends to apply the requirements. If the contract calls for the contractor to provide services to riders of the contract, that option can be included here, identifying each and all riders.

A brief summary of the JO authorization process is typically included. The owner may require a *non-execution clause* advising contractors that they must respond to a JO proposal request or accomplish JOs authorized during the term of the contract, provided these JOs are in compliance with the contract's scope and intent. If the contractor believes the request was not in compliance with the scope and intent of the contract, a written request to decline must be submitted to the owner. This request should describe how the owner's request fails to comply with the contract.

Construction standards and/or technical specifications can be referenced as applicable if not already sufficiently addressed in other sections. These are owner-developed or industry standards and reflect the owner's requirements for standardized use and application of various materials, equipment, and products.

## Contract Terms & Dollar Amounts

Most JOC contract terms provide sufficient duration for the contract to reach its full potential for both the owner and the contractor, which is typical protocol for contracts with ID/IQ features. Contractors often want to spread start-up costs over a longer period of time, and owners want to avoid the effort and expense of re-bidding contracts.

Owners generally include a guaranteed minimum amount of work to be authorized during the contract term. Maximum annual aggregate or individual total amounts of JOs accomplished and paid per contract term are sometimes imposed by governing bodies.

### Terms

This section of the General Requirements describes the term, or duration, of the contract. Contract terms are usually one base year with options for annual renewal, subject to bilateral agreement for contract extension between both parties to the contract, without providing justification for non-renewal by either party. If option years are included, most contracts allow two to four years, potentially renewable each year. The owner provides a written notice to the contractor with intent to renew the contract—usually 30 days in advance of the end of the contract term. Job orders can be authorized at any time during the term of the contract.

There is a current trend toward allowing owners the option to award multiple JOC contracts simultaneously to multiple contractors. If the owner is contemplating single award, some contracts include a provision that allows the owner to award a second contract at any time during the first six months of the term. If the owner's purchasing requirements allow this, the next highest ranked respondent from the original RFP evaluation process may be offered an award, without having to re-solicit the RFP. In this case, the post-awardee's original price proposal would not be altered.

### Dollar Amounts

This section specifies the dollar amount authorized for payment during the term of the contract. A *minimum guaranteed annual value* of $50,000 or more is sometimes established as a legal "consideration." However, not all contracts identify a guaranteed minimum annual value, and many contractors may feel this is of little relevance, since, typically, the guaranteed amount is much less than the cost of preparing a proposal and mobilizing for the contract. Successful JOC contractors are more concerned about the maximum volume they may obtain if they become the owner's "preferred provider" of services.

Contracts do not guarantee any annual values to the contractor above a stated minimum. This feature allows the owner the option to curtail requests for JO proposals at any time during the term of the contract, once any minimum annual value requirements have been satisfied. Governing bodies may require contracts to identify at least some minimal guaranteed commitment by both parties—an issue that should be researched prior to contract development. Each individual JO typically has a minimum value of $1,000-$2,000, sometimes as much as $5,000, or no required minimum at all.

Contracts sometimes identify a *maximum total annual value*, usually up to the low millions, based on the owner's anticipated volume of projects earmarked for JOC. However, most contracts identify no maximum total annual value, offering flexibility to accommodate increasing volumes of work. Individual JOs may or may not be restricted to maximum values. Some governing bodies have restricted individual JOs to a maximum of $750K, others up to $1M. The federal limit is $3M. Other governing bodies (such as Texas, at the state level) have no restriction at all.

## Definitions

This section of the General Requirements defines terms used in the contract—if not already sufficiently defined in other sections or if there is a need to amend previous definitions. Those specific to JOC may include, but are not limited to, the following:

- Job Order Contract
- Contract Manager
- Project Coordinator or Administrator
- Indefinite Delivery/Indefinite Quantity
- Job Order
- Unit-Price Book
- Non-Prepriced Work
- Coefficients
- City Cost Index Multipliers
- Standard and Non-Standard Hours
- Emergency Work
- Code Review Drawings
- Maintenance Work (if applicable)

## Program Management Interface

This section of the General Requirements references the designated UPB or group of UPBs with specific software applications. It addresses hardware and software applications needed to interface data and communications between the contractor and the owner. Included within this section is any formal training the contractor must provide to the owner's key contract management staff—usually as soon as possible after contract award.

## Professional Consultants

This section of the General Requirements covers professional architectural and engineering (A/E) services, including consulting and/or design efforts, which may be determined by the owner's jurisdictional purchasing requirements.

### Architectural & Engineering Services

When professional A/E services are required or desired for the project under consideration, owners can engage directly in separate contractual agreements for these services, use the owner's in-house designers, or obtain these services through the JOC contractor. During the term of a large JOC, all three methods may be employed at various times. JOC contractors might have professional consultants on staff, readily available to provide services as needed, or they may have a local A/E firm as a member of the team.

If professional consulting services cannot be obtained through the JOC contractor and are needed frequently, the owner might consider using separate ID/IQ agreements for A/E services to specifically accommodate the JOC program. In any case, the owner should promote designer-contractor relationships as early as possible. JOC offers a desirable opportunity for professional consultants, whether they contractually engage directly with the owner or through the JOC contractor. If a third party A/E is used, the owner must ensure that the A/E understands the JOC program, the fact that 100% drawings are often not required, and that response time is critical.

With JOC contracts, the contractor usually obtains required permits on the owner's behalf (unless restricted from doing so). Often, stamped and sealed documents generated by a design professional are mandatory in order to obtain permits from various regulatory bodies. These are usually municipalities or other federal or public-sector entities that have jurisdictional authority to permit and inspect work for compliance with applicable codes, statutes, standards, or regulations. For example, the local fire department might serve as the authority having jurisdiction (AHJ) over specific details of project design that affect the health and life safety of the building occupants, such as emergency access to or from buildings.

Other permitting entities may include city building inspectors or state-sanctioned inspectors authorized to ensure requirements are met regarding issues such as ADA compliance, historic building preservation, and environmental protection. If this is the first time that JOC has been used in the permitting entity's jurisdiction, an open house and briefing by the JOC team (both owner and contractor) on the process and team capabilities helps facilitate understanding of the unique JOC process and tends to expedite permit approval.

Professional consultants ensure that new construction or modifications to existing structures comply with current codes and statutes, such as the NFPA 101 *Life Safety Code*, the *International Codes*, the *National Electric Code (NEC)*, the *Americans with Disabilities Act (ADA)*, etc. Design professionals can verify that new construction and modifications to existing structures are designed to be structurally safe for occupants, and that other types of projects meet currently recognized design criteria to protect facility users and the environment. Consultants can also enhance the functionality and maintainability of the facility's physical components and help ensure that construction standards and unity of design specific to the facility are met.

A professional consultant's design efforts in support of a JOC program can be less than what is required for other delivery methods. Consequently, compensation for these types of services may be comparatively less per project than design efforts that require comprehensive and complete design documents, as needed for project-specific competitive bidding. However, JOC interface may prove worthwhile to the designer, depending on the volume of work offered.

Quick turnaround by the consultant via an ID/IQ contract, using standard specifications and design details, can reduce pre-construction time. Compensation for architectural work may be based on a negotiated price per square foot for interior design services or on negotiated hourly rates. If professional design services are subcontracted by the JOC contractor, a fixed percentage of project construction costs for design services can be negotiated in accordance with contract provisions, or negotiated at the time of JOC contract award.

Professional design fees can be an allowable line item or, in some cases, identified as a contractor's responsibility and included in the coefficient. Normally, these services must be obtained based on qualifications rather than price. The procurement of professional design services should always be in compliance with any applicable requirements by governing bodies.

Long-term relationships among designers, owners, and contractors are not new in the industry and have a long history in the private sector.

JOC practice in the public sector can accommodate and enhance these relationships. With JOC, the design process becomes more efficient as the designer becomes familiar with client expectations. There is increasing opportunity for design professionals to participate in JOC—to fully *partner* in the process.

## Price & Payment Procedures

JOC is often referred to as "open book" in regards to JO pricing. This, in part, is due to the distinct way all pricing components are openly displayed and accounted for at all times, from the time the contract is awarded through the authorization and execution processes. It is necessary to understand pricing requirements for JOC, including JOs, change orders, and emergency work.

### Job Order Pricing

This section of the General Requirements identifies primary components associated with JO pricing. Pricing may vary with each component to meet owner or co-op requirements. Co-op pricing is different from owner-awarded pricing, as members' individual requirements for projects to be accomplished can be intermittent (especially those of smaller members of the co-op), and sometimes spread out over large geographic areas away from the contractor's base operations.

Once established, the primary components of JO pricing are used as the criteria for establishing the sum total of UPB line items identified to reflect the scope of work. This sum is adjusted by the contractor's applicable coefficient (and any other designated multipliers) to determine the price of the JO. Multiple UPBs may be designated to allow for additional pre-priced items, and if so, the least expensive like-item should prevail unless the owner requests a specific, more expensive item. *(See "Non-Prepriced Work" later in this chapter.)*

---

**Price & Payment Procedures**

- Job Order Pricing
  - UPB Costs Column
  - City Cost Index Multipliers
  - Division 1 Costs
  - Non-prepriced Work
- Change Order Pricing
- Emergency Work Pricing

- Overtime
- Contingencies
- Coefficient Adjustments
- Invoicing & Payment

---

**Figure 5.2** identifies sections and components of the General Requirements that are associated with pricing and payment procedures.

## UPB Costs Column

The sum total of UPB line items for JO pricing is derived from the designated costs column, e.g., UPB "Bare Costs - Total" column or "Total Incl. O&P," as designated in RSMeans cost data. These costs can be adjusted by the designated city cost index multipliers (if applicable). Note that the costs columns reflect national averages for the installing contractor (subcontractor) but not for the general or prime contractor.

Costs column designation can make for interesting discussion among JOC practitioners—especially regarding a contractor's in-house workforce—as the "Total Incl. O&P" costs column intermingles with some Division 1 line items. For instance, Total O&P pricing for line items includes profit and overhead costs for the installing contractor. Division 1 line items also include overhead expenses. They usually are aligned with general contractor costs but will apply to subcontractors under some circumstances. Since the allowance of certain Division 1 items affects the contractor's overall O&P per project, careful consideration should be given to which items should be allowed or restricted. Designation of either column in the contract will be reflected in the contractor's competitively bid coefficients. *(See "Unit Price and Coefficients" in Chapter 2 and "Coefficient Development" in Chapter 6 for more information.)*

## City Cost Index Multipliers

RSMeans' UPB line items reflect national averages for material and installation costs. City cost index multipliers provide a way to adjust these costs to the owner's regional area of operations, using percentage ratios. UPB software applications can automatically adjust line items to designated city cost index multipliers. Not all contracts require use of a city cost index multiplier. If not, the coefficient bid by the contractor will reflect this.

JO contracts identify the extent of allowable UPB line items. In most cases, items in other divisions are allowable unless the owner restricts them. For Division 1, however, allowable or restricted items are a major issue regarding JO pricing. They directly affect the contractor's development of the coefficient—and therefore directly affect the cost of each JO. The designation of any line item as allowed or restricted should be clearly indicated in the contract documents.

## Division 1 Costs

Division 1, General Requirements, identifies subdivisions relating to administrative and quality requirements, temporary facilities, controls, and so forth. These can significantly impact the cost of a project—especially cost adjustment factors for repair and remodeling work (as UPB costs are based on new construction), modifications for project job conditions, markups for subcontracted services, and overtime. Other

Division 1 line items include costs that are not necessarily project-specific, such as those associated with the contractor's main office overhead, insurance, bonding, field personnel, etc.

Depending on contract requirements, JO pricing structures fall into one of the following three scenarios:

- Never allow the use of Division 1 line items.
- Allow the use of some Division 1 line items.
- Allow the use of all Division 1 line items.

The owner considers the JOC pricing structure (in relation to the General Requirements division of the designated UPB) while the requirements for services to be provided by the JOC contractor are being assessed. The owner may decide that a substantial restriction of Division 1 line items will expedite the JO price proposal approval process by eliminating the need to review and consider the applicability of allowable Division 1 costs as they pertain to each project. (This means they are to be included in the contractor's coefficient.) This issue is addressed as a component of JOC implementation strategy and, subsequently, reflected as a contract requirement.

Figure 5.3 is a summary of Division 1, General Requirements, from RSMeans, including designated UPB line item allowances. It serves as an example only and reflects a contract allowing only some Division 1 line items. Other items are to be included within the contractor's coefficients or may not be allowed at all as part of the contract.

### Non-Prepriced Work

Non-prepriced work can be defined as tasks not covered by the UPB but by the contract's scope and general intent. The cost of these items may be negotiated by the owner and incorporated into the price of the JO. Contracts may vary slightly on requirements pertaining to non-prepriced items.

Some contracts limit the use of non-prepriced items to a maximum of 10% of the total number of line items within an individual JO price proposal. The following methods are among those used to estimate non-prepriced items:

1. *Cost Plus Fee Method.* This method may be used when unit pricing has not been pre-established for tasks such as fabrication of specialty items, the use of very specialized subcontracted services, or uncommon materials or devices specified by the owner. Compensation to the contractor can be based on the cost of the specialty service or item plus a fee (usually negotiated at approximately cost plus 10%) or by a separate coefficient. When possible (and based on availability), the owner may request

## Allowable Division 1 Items for JO Pricing (Example)

| | (A) Allowed as Directed in JO | (B) Not Allowed/ Included in Coefficient | (C) Not in Contract |
|---|---|---|---|
| **01100 Summary** | | | |
| **01103 Models & Renderings** | | | |
| 200 Models | X | | |
| 500 Renderings | X | | |
| **01107 Professional Consultant** | | | |
| 100 Architectural Fees | X | (See General Requirements for JOC) | |
| 200 Construction Management Fees | | X | |
| 300 Engineering Fees | X | | |
| 700 Surveying | X | | |
| **01200 Price & Payment Procedures** | | | |
| **01250 Contract Modification Procedures** | | | |
| 200 Contingencies | X | | |
| 400 Factors | | X | |
| 500 Job Conditions | | X | |
| 600 Overtime | | X | |
| **01255 Cost Indexes** | | | |
| 400 Historical Cost Indexes | | | X |
| City Cost Indexes | | X | |
| **01290 Payment Procedures** | | | |
| 800 Taxes | State Sales Tax Exempt | | X |
| **01300 Administrative Requirements** | | | |
| **01310 Project Management/Coordination** | | | |
| 150 Permits | X | | |
| 200 Performance Bond | | X | |
| 300 Construction Time | | X | |
| 350 Insurance | | X | |
| 400 Main Office Expense | | X | |
| 500 Markup | | X | |
| 610 Overhead | | X | |
| 700 Field Personnel | | X | |
| **01320 Const. Progress Documentation** | | | |
| 200 Scheduling | | X | |
| **01321 Construction Photos** | | | |
| 500 Photographs | X | | |
| **01400 Quality Requirements** | | | |
| **01450 Quality Control** | | | |
| 500 Testing | X | | |
| **01500 Temporary Facilities & Controls** | | | |
| **01510 Temporary Utilities** | | | |
| 800 Temporary Utilities | | X | |
| **01520 Construction Facilities** | | | |
| 500 Office | | X | |
| 550 Field Office Expense | | X | |
| 900 Weather Station | | | X |
| **01530 Temporary Construction** | | | |
| 700 Protection | X | | |
| 900 Winter Protection | X | | |
| **01540 Construction Aids** | | | |
| 500 Personnel Protective Equipment | | X | |
| 550 Pump Staging | X | | |
| 700 Safety Nets | X | | |
| 750 Scaffolding | X | | |
| 755 Scaffolding Specialties | X | | |
| 760 Staging Aids | X | | |
| 780 Swing Staging | X | | |
| 790 Surveyor Stakes | X | | |
| 800 Tarpaulins | | X | |

**Figure 5.3** illustrates select General Requirements items in relation to a JOC contract's job order pricing structure. It provides an example of how each line item can be used. Modifications can be made as needed to meet each individual owner's requirements. Column A identifies line items allowed only with the owner's permission. Column B identifies line items that are not allowed in JOs. The costs associated with these line items are to be included in the contractor's coefficients. Column C identifies line items that are excluded from the JOC contract.

| | (A) Allowed as Directed in JO | (B) Not Allowed/ Included in Coefficient | (C) Not in Contract |
|---|---|---|---|
| **01500 Temporary Facilities & Controls** | | | |
| **01540 Construction Aids** | | | |
| 820 Small Tools | | X | |
| **01550 Vehicular Access & Parking** | | | |
| 700 Roads & Sidewalks | X | | |
| **01560 Barriers & Enclosures** | | | |
| 100 Barricades | | X | |
| 250 Temporary Fencing | X | | |
| 400 Temporary Construction | X | | |
| 800 Watchman | | | X |
| **01580 Project Signs** | | | |
| 700 Signs | X | | |
| **01590 Equipment Rental** | | | |
| 100 Concrete Equipment Rental | X | | |
| 200 Earthwork Equipment Rental | X | | |
| 400 General Equipment Rental | X | | |
| 500 Highway Equipment Rental | X | | |
| 600 Lifting & Hoisting Equipment Rental | X | | |
| 700 Wellpoint Equipment Rental | X | | |
| 800 Marine Equipment Rental | X | | |
| **01740 Cleaning** | | | |
| 500 Cleaning Up | | X | |
| **01810 Commissioning** | | | |
| 100 Commissioning | X | | |
| **01830 Operation & Maintenance** | | | |
| 500 Facilities Maintenance Equipment | X | | |
| **01832 Facilities Maintenance** | | | |
| 220 Site Work Facilities Maintenance | X | | |
| 230 Concrete Facilities Maintenance | X | | |
| 240 Masonry Facilities Maintenance | X | | |
| 250 Metals Facilities Maintenance | X | | |
| 270 Moisture-Thermal Control Facilities Maintenance | X | | |
| 280 Door & Window Facilities Maintenance | X | | |
| 290 Finishes Facilities Maintenance | X | | |
| 300 Specialties Facilities Maintenance | X | | |
| 310 Architectural Equipment Facilities Maintenance | X | | |
| 320 Furnishings Facilities Maintenance | X | | |
| 340 Conveying Systems Facilities Maintenance | X | | |
| 350 Mechanical Facilities Maintenance | X | | |
| 360 Electrical Facilities Maintenance | X | | |
| **01840 Moving Equipment** | | | |
| 100 Moving Equipment | X | | |
| **Divisions 2 Through 16** | | | |
| Divisions 2 Through 16 | X | | |

**Figure 5.3** (continued)

three written quotes from the contractor, including any contractor discounts, to document good-faith competitive pricing for these types of tasks or materials.

2. *Like Item Method.* This method specifies a UPB line item that is basically the same in material function and installation as the non-prepriced line item.

Previously non-prepriced items are identified in the UPB breakdown of the JO price proposal with designated auditable documentation. Once approved, they can be permitted as directed, becoming pre-priced items for the term of the contract.

## Change Order Pricing

Change orders are written modifications to the scope of work of a JO, after authorization to proceed. Like the JO proposal, a scope of work for the change is agreed on and followed by an owner-approved price proposal for the change in accordance with JO pricing requirements. The result is a net decrease or increase in project costs. Most JOC contracts handle change order pricing by adjusting the quantities of UPB line items already identified—if the change is authorized before the affected trades are mobilized. A standard change order form can be used to reflect the change, describing or referencing the written change in the scope of work and pricing with an attached UPB breakdown. Alternatively, a revised JO can be authorized, reflecting the change in scope, the revised UPB breakdown, and the revised firm fixed price. For situations in which remobilization will be required, the owner can include a reasonable added percentage to the costs for the change as designated in the contract.

## Emergency Work Pricing

Some JOC contracts unilaterally designate a differential increase to JO pricing (usually 25%) for non-warranty emergency work-in-place. This work may be performed during standard hours (usually 7 a.m. to 5 p.m., Monday through Friday) or non-standard hours. Emergency work pricing applies to the 24-hour period immediately following a request to the contractor to allocate, as a top priority, all resources necessary to accomplish work on an emergency basis. After the 24-hour period, if work is still in progress, standard or non-standard hour pricing is used for pricing the remainder of work-in-place, as applicable. Owner requirements may vary.

## Overtime

JOC contracts typically do not allow overtime payments, which should be included in the contractor's coefficients. The costs for overtime to meet agreed-on JO completion targets must be borne by the contractor. If the scope of work changes, extensions to the time line of a project should be allowed as appropriate. If time cannot be extended and changes are necessary, additional shifts may be required, using both standard and non-standard work performance times with respective coefficients as applicable. Owner requirements may vary.

## Contingencies

Renovation, alteration, and rehabilitation projects are inherently susceptible to unforeseen field conditions and other factors such as owner-requested changes that can increase the cost of the project. With JOC, only changes in the scope of work constitute the issuance of change orders. Construction contingency allowances (usually 8%-10%

of the JO sum total amount) are commonly applied to these types of projects and with the cost for changes deducted from the contingency amount in accordance with change order pricing requirements.

## Coefficient Adjustments

Most owner-awarded contracts do not allow coefficient adjustment during the base term or option years of the contract. To stay in sync with fluctuations in material costs and labor rates, the owner and the contractor should use the most current edition of the designated UPB database (with downloadable updates) for the base term and for any subsequent option terms. This method, in most instances, provides fair compensation to the contractor for any option years awarded, without adjustment to the original coefficient proposed at the time of contract award.

However, some contracts identify the most current edition(s) of the designated UPB(s) at time of award as the only edition to be used for the term of the contract, even if option years are awarded. In this case, adjustments to the contractor's original coefficient are warranted and a recognized historical cost index, such as RSMeans *Construction Cost Indexes* or *Engineering News-Record,* can serve as a method to adjust original coefficients bid for option years awarded. These index figures are based on national labor and materials cost fluctuations and market trends in the construction industry. The index figure in place at the time of award, as well as the figure established for each option year, is used to adjust the coefficient established at the time of contract award.

Contracts awarded by purchasing cooperatives may include unilaterally applied coefficient adjustment multipliers. In these cases, contract provisions allow pre-established positive or negative multipliers applied to contractor-established coefficients based on certain circumstances:

- Travel distance from the contractor's base point of operations
- Increases or discounts tied to the relative value of each JO
- Standard or non-standard work performance times
- Discounts tied to the annual value of work authorized by the owner

This concept is innovative and is currently in practice through cooperative-awarded JOC contracts.

## Invoicing & Payment

Payment procedures can include the owner's usual requirements for invoicing, procedures for progress payments, and final payment after the work is accomplished and approved. It is important that all required close-out documents be received and approved by the owner prior to final payment.

## Summary of Job Order Pricing & Payment

As with any delivery method, establishing clear requirements for pricing and payment are essential. This is especially true with JOC, since JO pricing structures are distinctly different from those of other delivery methods. Clarity in the contract is also important to reduce the potential for disputes and to expedite the JO and change order authorization processes.

To summarize the primary pricing components of the contract:

1. The contract identifies the following:
   - Designated UPB and software applications
   - UPB costs column
   - City cost index multiplier (if applicable)
   - Allowable UPB line items from Division 1 and the entire UPB
   - Methods to address the pricing of non-prepriced work, emergency work, and changes to the scope of work
2. The contractor establishes a coefficient for each wage rate and hours of performance category identified by the owner.

## Administrative Requirements

Administrative requirements include contractor responsibilities for managing, coordinating, and executing projects; attending meetings; and identifying documents that must be submitted to the owner. Other administrative requirements include project scheduling and status reports of requested and active JOs.

---

**Administrative Requirements**

**Project Management/Coordination**
Job Order Authorization
- Request for JO Proposal
- Site Visit
- Scoping
- JO Proposal Submittals
- JO Price Proposal

**Contractor's Requirements**
Meetings
- Program Start-up Meeting
- Program Progress Meetings

Submittals
Shop Drawings
Cutting & Patching
Regulatory Requirements
- Codes & Permits
- Code Review Drawings
Special Project Procedures
Field Personnel

**Construction Progress Documentation**
Status Reports
Scheduling

---

**Figure 5.4** lists contractor administrative requirements. The JO authorization process is unique and is a primary component in JOC's ability to reduce pre-construction time lines.

## Project Management/Coordination

This section of the general requirements covers project management, coordination, and administration. The JO authorization process is unique to JOC, exhibiting the method's primary aspect of partnering. A thorough program start-up meeting is particularly important and should take place as soon as possible after the contract is awarded. This should be followed by periodic progress meetings.

The contractor may be required to develop and submit drawings reflecting the project's compliance with applicable codes. Most other items in the Project Management/Coordination section are somewhat generic and similar to those in the contract documents of other delivery methods.

### *Job Order Authorization*

This process consists of four main procedures, beginning with the owner's request for a JO proposal from the contractor. The JO proposal request should include a brief description of the project, location, and desired or required completion date. The owner and contractor (and other key participants, if necessary) then visit the site to "scope" the project. The results of scoping provide information needed for the third procedure—the development of drawings (if necessary); a concise, written scope of work; and a detailed price proposal by the contractor—the combination of which is the JO proposal.

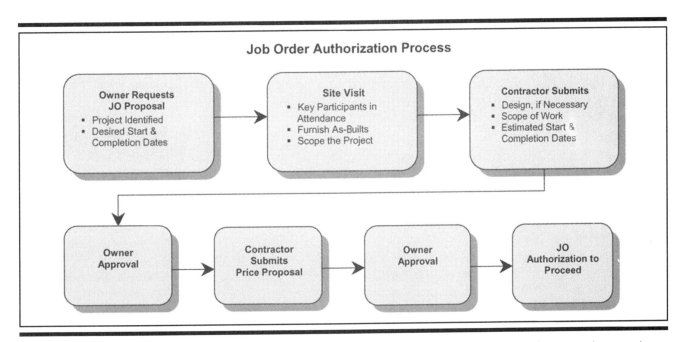

**Figure 5.5** illustrates the JO authorization process. Note that the design, scope of work, and the project's estimated start and completion dates are approved by the owner before the contractor's price proposal is submitted for owner approval.

Lastly, the JO proposal is submitted to the owner in a two-step approval process resulting in authorization to the contractor to accomplish the work of the project. This process is a primary reason JOC makes it possible to accelerate pre-construction time lines.

## Request for JO Proposal

The process for the owner to initiate work under a JOC contract begins with the owner's issuance of a written Request for a Job Order Proposal to the contractor. The time frame required for the contractor to respond to the request can vary. However, most contracts require the contractor to respond within two to four working days after the date of request, unless designated as an emergency by the owner.

The request contains a brief description of the project and its location. This generally includes desired or required start and completion dates and contact information for scheduling a site visit or "job walk" with the owner's designated representative. Other information, such as work performance times, professional consulting services, code review drawings, and permits, can also be noted. The request should be concise but contain enough information for the contractor to preliminarily understand the project and be prepared for the site visit.

### Multiple JO Proposal Requests

The issue of owners requesting multiple JO proposals for the same project has been a topic of much discussion among JOC practitioners. Owners often request competitive pricing for contract award purposes utilizing project-specific delivery methods. There are well-known ethical practices for these situations by which every owner and contractor should abide—such as prohibiting the sharing of contractor pricing information among potential awardees prior to award in an effort to save money (a practice known as *bid shopping*).

With JOC, the argument can be made that the contract has already been competitively bid and awarded. Therefore, the practice for owners to competitively bid each JO among multiple awardees and/or from available co-op awardees is in conflict with the core principles of the JOC method and detrimental to building relationships. Competitive bidding at this level is considered unethical among some practitioners.

However, if a single JO request is issued and the JO price proposal approval process becomes stalled after reasonable negotiation efforts have been exhausted, most practitioners believe informing the initial proposer of intent to re-issue the request to another available awardee is warranted.

## Site Visit

The contractor acknowledges the request for a JO proposal by contacting the owner's designated representative to arrange for a site visit within the required time period. Key stakeholders might also be in attendance. These may include professional consultants (if necessary), facility operations staff, subcontractors, or "end-users" if programming has not already been completed. Their input can be beneficial in the scoping process.

The owner should arrange to provide access to all areas affected by the project so that key participants in the scoping process can examine existing conditions. A concise written record of the site visit, to include sketches, is typically developed by the contractor and/or designer (if necessary) during the site visit and initialed by all parties. This helps to define and document the desired scope of the work. However, the documents generated at the site visit may represent only a defined *preliminary* scope of work, subject to subsequent review by others and/or receipt by the owner in a final format as designated by contract requirements. This should be noted when initialing these documents at the site.

## Scoping

Accurate scoping is important to identify all tasks required to accomplish the work, including job conditions, construction aids, equipment, and material applications. JOC contractors know that changes in the scope caused by inaccuracies can pose financial liabilities to them—since a major component in JO pricing is identification and verification of line items required to perform the project's scope of work.

A successful JOC contractor will have highly trained, effective project management staff present during a project site visit. Contractors can assist the owner and designer with value engineering suggestions, technical assistance, and constructability input during the site visit. Special consideration should be given to the impact the project will have on the following:

- Existing site characteristics and features inclusive of systems and operations
- Any applicable codes, statutes, or regulations
- The possibility of disturbing hazardous materials
- Environmental protection—both indoor and outdoor, as applicable

If the project is to take place within occupied buildings, control measures to protect occupants from potentially adverse conditions, especially those associated with fire detection or suppression systems, fire egress, and indoor air quality, should be addressed. Owners can assist contractors in the scoping process by accommodating contractor requirements for staging, access, temporary facilities, and controls, and by furnishing all the information they can about the project such as existing site characteristics.

Owner-furnished record drawings ("as-builts") can be useful, since they reflect existing site characteristics. (These are, of course, of greatest value if they have been sufficiently updated.) If available, these drawings should be brought to the site visit to assist in scoping, as appropriate. A few copies of existing schematic drawings encompassing the limits of construction, building floor plans, MEP documentation, and utility maps are always useful for site visit attendees to become familiar with site characteristics.

All drawings should be proportionately scaled, and measurements field-verified. The contractor should have open access to the owner's as-built documents, as needed, during or after the scoping process. Digital photography can also be useful when documenting existing site conditions. In fact, it is often an essential step in developing an accurate scope.

The project's performance times should be addressed in the context of the owner's operations, as should the contractor's own issues including contractor mobilization, coordination, scheduling of trades, site protection and security, and any other items commonly addressed at pre-construction meetings.

### Construction Aids & Equipment

With JOC, it is important during scoping to address the construction aids and equipment that the contractor expects to use. Especially important are items to be used for accessing work surfaces, staging and placing equipment and materials, and excavation.

The *means and methods* used to perform work are traditionally at the contractor's discretion, although sometimes they are specified by professional consultants. Usually they are not of major significance to owners when using other delivery methods. Owners do not want the potential for increased liability that may be incurred by requiring specific means and methods. However, the owner's knowledge of how work will be performed by the contractor can assist in the JO approval process if the owner already anticipates the inclusion of allowable construction aids and equipment line items in the JO price proposal.

Arguably, since the JO price proposal becomes a lump-sum fixed price for the agreed scope of work, the owner should not be overly concerned with means and methods if the inclusion of these types of line items in the JO price proposal (construction aids and equipment) are proven accurate when compared to actual use and work is accomplished timely and safely in accordance with applicable codes.

## JO Proposal Submittals: Design, Scope of Work, & Performance Times

Once the site visit is concluded, the contractor initiates the development of design documents (if necessary). This may include code review drawings by the JOC contractor or A/E design through the contractor, accompanied by a detailed scope of work. The design and the scope of work, together with the project's estimated start and completion dates as desired or required by the owner, are the initial components of the JO proposal.

The scope of work should describe all the tasks needed to accomplish the work of the project. Standardized forms can be developed that categorize the project's identified task descriptions by CSI MasterFormat division. This provides all reviewers with a consistent format, assists in reducing errors and omissions, supports cost estimating efforts, and can expedite the JO proposal approval process. Often, if the project's scope is relatively small, the UPB line item breakdown can adequately document it.

The scope of work is then submitted as a draft to the owner for review and comments. Together with code review drawings (if necessary), the scope of work should contain enough information for the contractor (and owner) to estimate the project costs.

Sometimes, the owner may ask the contractor to submit a "ballpark" cost estimate, along with a draft scope of work, to be used for budget assistance and consideration of possible value engineering options.

Once the draft scope of work is received, it can be reviewed by owner representatives in accordance with the review policies and procedures the owner has adopted. In some cases, this may require allotting sufficient time for input from various stakeholders, including the owner's facilities maintenance and operations staff, as they review the scope of work and any related drawings and specifications.

Any identified changes or modifications are submitted to the contractor, and the end result is a final draft issued to the owner for consideration. Once the owner approves the design (if necessary), the scope of work, and the dates and times of performance, the contractor can proceed with developing the JO price proposal (which most likely has already been in development by the contractor, concurrent with the development of the scope of work).

### Response Times for JO Proposals

Contractors typically respond to a request for a JO proposal within a few days, in accordance with contract provisions. However, after the site visit, there are numerous factors that can affect the time line for submittal of initial JO proposal components. Figure 5.6 illustrates goals established at a JOC partnering meeting regarding contractor response times to submit the initial components of a JO proposal.

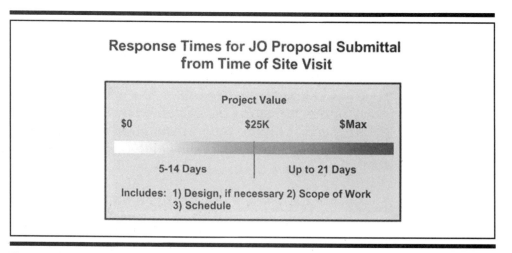

**Figure 5.6**

Response times vary with the complexity of the project. The owner should take all factors into consideration and allow reasonable expectations relative to each project. The average time from JO request to JO approval is 10–15 days, subject, in part, to the owner's decision-making process.

### JO Price Proposal

The JO price proposal is the final component of the JO proposal to be submitted to the owner for consideration and approval. It should contain an accurate, detailed listing of qualified and quantified UPB line items, reflecting the scope of work. The sum total of the UPB line items designated to accomplish the work of the project should clearly list each item, consistent with JO pricing requirements. This includes city cost index (if required), costs column, the contractor's applicable coefficient, and any non-prepriced items. Any deviation from the content of allowable line items should be either prohibited or owner-approved in accordance with contract requirements and clearly noted with support documentation.

Many owners develop their own cost estimate prior to, or concurrent with, the contractor, using the same pricing requirements. The owner's version can be used for comparison to verify the accuracy of the contractor's price proposal if the owner is sufficiently staffed for the task. If not, a review of the price proposal for accuracy can be done by transferring line items into the owner's cost-estimating application and reviewing all proposed line items to verify allowable line items for pricing accuracy and reasonable applicability to the agreed-on scope of work. Note that every line item may not have an exact match to a project.

Owners will not be successful administrators of a JOC program if every project is negotiated with the intent to provide an absolute minimum compensation to the contractor. Similarly, contractors will be unsuccessful if they consistently try to "nickel and dime" the owner. The scope of work and price proposal development and approval process can be expedited if the owner and the contractor work in concert—partnering throughout both processes while utilizing designated management interface resources to the fullest extent possible.

## Contractor Requirements

Contractor requirements commonly identified within other types of contracts used for renovation, alteration, rejuvenation, or upgrade projects can also be used for JOC. These requirements usually include coordinating the work of subcontractors, assuming responsibility for protection in and around the project site, site security, maintaining regulatory and project documents at the site, noise and dust control, and any other special owner requirements. Unless line items involving these tasks are allowable, they are to be included in the contractor's coefficient(s).

### Meetings

This section of the General Requirements addresses start-up and follow-up progress meetings. As defined in earlier chapters of this book, the start-up meeting sets the stage for communication and relationship building among key participants. This is an essential ingredient in the success of a JOC program, and participants must be aware of one another's policies, procedures, and expectations in order to reach the full potential of the delivery method.

#### Program Start-up Meeting

As mentioned earlier, a program start-up meeting is advised immediately after contract award. This will help establish mutually beneficial relationships among key participants. Attendance should be mandatory for all key participants and stakeholders, with representation from the owner, the contractor, and, if possible, the designer.

The meeting should be presided over by the owner's JOC CM. Typically, one full day should be committed for this meeting. If possible, it should take place in a formal setting, such as a training environment. In some instances, the inclusion of an experienced, independent facilitator is helpful, especially when deemed appropriate by either the CM or the JOC contractor.

Attendees representing the owner should include the following:

- JOC CM
- JOC contract administrator(s)
- Estimator(s)
- Inspector(s)
- Designer(s) (in-house or otherwise)

Optional owner representatives may include the following:

- Facilities maintenance supervisor(s) or a designated representative
- In-house trades supervisor(s) or a designated representative
- HUB coordinator(s)
- Representative(s) from purchasing
- Representative(s) from risk management and/or environmental health and safety
- Key administrative assistant(s)

Attendees representing the contractor should include the following:

- Contractor or contractor's representative designated as a single point of responsibility to the owner for contract compliance
- Project manager and superintendents(s)
- Quality-control inspector
- Designer(s) (in-house or otherwise)

Optional contractor representatives may include the following:

- Contractor executive(s)
- Major subcontractors(s)
- Key administrative assistant(s)
- Accountant(s)
- Other key support staff

The owner's in-house design team should attend, if applicable. If the owner or the contractor has an ID/IQ contract in place with a design professional, that person should also attend. It is not imperative that the designer attends, but it is beneficial. If the designer or design team cannot attend, the requirements for professional consulting services (if applicable) and/or code review drawings by the contractor should be addressed.

Agendas for program start-up meetings are written by the owner and distributed in advance to all attendees within a reasonable time frame (one week minimum), allowing time to gather any forms and documents that may be useful. The following topics are presented as an example of agenda items that may be addressed at the start-up meeting:

1. *Identification and contact information for all attendees.* Attendees should briefly explain their respective designated roles as program participants or stakeholders, their responsibilities, and their levels of authority as they relate to JOC processes. This sets the stage for communication.

2. *Identification of mutual program goals.* A reasonable amount of time is allotted at the meeting for all attendees to voice their expectations. Small groups charged with identifying mutual goals may be formed. Goals defined by each group can then be posted and addressed. Any apprehensions associated with identifying mutual goals should openly be addressed at this time. There should be a consensus to commit to the goals of the program, as well as to the basic principles of partnering.

3. *Charting of program processes and procedures.* All owner processes associated with project planning, establishing budgets, work authorizations, change orders, contract administration, invoicing requirements, and payment should be addressed. All contractor processes associated with program start-up, use of owner's premises, JO proposal development (including code review drawings, scope of work, and price proposals), and project management/coordination are explained. Other topics include processes for submittals, shop drawings, regulatory requirements, codes and permitting, temporary facilities and controls, quality control and inspections, cleaning, rubbish handling, and JO closeout requirements. Any HUB requirements and record retention processes are addressed. There should be a consensus of agreement to the processes.

4. *Identification of all standard forms, documents, and data that may be exchanged between key program participants.* This includes any applicable standardized general technical specifications and any design criteria or materials specifications that are specific to the facility. If standardized groups of UPB line items for repetitive types of work can be mutually agreed on, this can be very useful for generating cost estimates and expediting the JO price proposal approval process.

5. *Scheduling processes.* Schedules and reporting requirements should be addressed, including when critical path method (CPM) or similar methods of scheduling are required.

6. *Contractor response issues.* Guidelines should be established for the contractor's: (1) responsiveness to voice/data communications, (2) appointments, (3) requests for JO proposals, (4) submittal of required documents, and (5) other response issues.

7. *Owner response issues.* Address the owner's responsibilities regarding responsiveness to: (1) voice/data communications, (2) appointments, (3) availability of designated owner's project

representative when needed, (4) approval of documents submitted by the contractor, and (5) other owner response issues.

8. *Quality performance goals.* Set mutually agreed-on performance goals for: (1) accuracy of estimating projects (such as +/- 10%), (2) starting and completing projects on time (90%), (3) starting and completing projects with critical time lines on time (100%), (4) the contractor's provision of technical expertise during the design phase, (5) keeping project costs within the established budget, (6) effective supervision, (7) adequate staffing of projects, (8) quality of workmanship, (9) minimizing changes, (10) minimizing punch lists, (11) HUB outreach and mentoring efforts, and (12) overall compliance with contract provisions.

9. *Identifying a means for informal dispute resolution.* Lack of effective communication is probably the single major cause for disputes. It can also be a reaction to (or symptom of) larger problems, real or perceived. The use of email, for example, works well for transferring data but is not a good vehicle for communicating disputed issues. Disputes are best addressed in person, whenever possible. Unless catastrophic in nature or of grave consequence, most dispute resolutions should be informal in nature and involve open communication and compromise among disputing parties. The owner's and the contractor's staffs should advise their supervisors of any problems promptly so the resolution process can be initiated swiftly.

## Program Progress Meetings

Subsequent annual program progress meetings are recommended, and bi-annual meetings should be considered for key contract execution participants. The agenda should focus on ways to streamline procedures already in practice. Agendas similar to the program start-up meeting may be needed if there is significant turnover of key participants in the program.

Progress meetings can be preceded by a semi-annual formal evaluation process, overseen by the owner's CM, to determine whether program goals are achieved, requirements are followed, and expectations are met. Standard forms can be developed for this purpose, listing basic performance requirements of the contract, commitments documented in the contractor's Management/Technical Proposal (submitted during the RFP process), and goals outlined at the program start-up meeting.

The contractor's performance should be evaluated by the owner's key representatives. Evaluation results can be submitted anonymously and summarized and shared with the JOC contractor. The contractor may be required to submit a written plan to remedy identified deficiencies. The contractor may choose to evaluate the owner's performance, as well, which can assist the owner in identifying correctable deficiencies.

Progress meetings are an opportunity to reflect on goals and ways to streamline processes. Time should be allotted for a Q&A session. Notes should be kept during the meeting, documenting the coverage of agenda items and any agreements, and should be distributed to meeting attendees.

The consensus to agree to the performance goals established during program start-up and progress meetings constitutes a contractual link to the JOC contract. Failure to sufficiently meet performance requirements and goals is a contract renewal issue for option years and, if it becomes a serious issue, could prompt the owner to stop issuing requests for JO proposals for the remainder the term of the contract. If this occurs, written notification by the owner in advance of pending action is appropriate.

Costs incurred for meetings can be shared equally by the owner and the contractor. Project-specific meetings involving progress scheduling or other issues should be held as deemed necessary by the owner or contactor.

*Pre-Construction Meetings*
Since most JOC projects are small in scope and scale, their project-specific pre-construction meetings usually take place at the project site and are referred to as site visits. *(See "Site Visit" in the "Job Order Authorization" section of this chapter.)* For JOCable projects of larger scope and scale, a more formal project-specific pre-construction meeting (including a site visit) may be necessary to make sure the project is planned properly. The necessity for this meeting is determined by the owner and the contractor. All relevant topics, including sequence schedule, staging, responsibility assignments, construction aids, performance times, protections, safety, and environmental issues, may be discussed. These types of project-specific pre-construction meetings are not to be confused with a JOC program start-up meeting, which is typically a one-time event geared toward establishing the initial partnering atmosphere and process among key program participants.

## Submittals
Submittal and approval requirements for mockups or field samples are no different for JOC than the owner's usual requirements for any other delivery method. However, due to JOC's potential for long-term relationships, a list of commonly used materials and devices specific to the owner's facility will help contractors reduce repetitive submittals and will assist in streamlining the JO approval process. Any non-prepriced items, as well as documentation of associated procurement requirements for such items (if applicable), are to be included in the submittal process.

### Shop Drawings

Shop drawings usually are defined as the diagrams, illustrations, schedules, performance charts, brochures, and other data, prepared by the contractor or any subcontractor, manufacturer, supplier, or distributor, that illustrate some portion of the work. The owner's established procedures for shop drawing submittal and approval apply. It helps streamline processes when required items to include shop drawings are identified by the owner in this section of the General Requirements.

### Cutting & Patching

This section addresses the owner's requirement for cutting and patching existing finishes. This section is similar to those used in other delivery methods for similar projects (renovations, alterations, rejuvenations, and upgrades to existing buildings and structures). The requirements for this section should be thorough and extensive since a large majority of JOCable projects include cutting and patching tasks.

### Regulatory Requirements

Any requirements for the contractor to obtain building permits, licenses, or inspections from all respective AHJs, on the owner's behalf, are addressed in this section of the General Requirements. Additionally, the contractor may need to submit code review drawings. This section may also list, as references, all statutes, codes, regulations, standards, and any other governing documents with which the contractor is required to comply regarding work to be performed under the contract, if these documents are not already listed elsewhere in the contract.

#### Codes & Permits

Applicable codes, standards, and regulations for governing work performed by the contractor are included in this section. Work should be in compliance with the latest editions of regulatory requirements unless otherwise specified by the owner or superseded by local AHJs. It is the contractor's responsibility to ensure that work performed complies with the regulations listed in this section. If professional consultant services are subcontracted by the contractor, their design efforts are to comply with current codes, standards, etc., and they may share in this responsibility with the contractor. *(See the "Professional Consultants" section of this chapter.)*

#### Code Review Drawings

This section includes the contractor's requirement to furnish drawings on each JO (if necessary) with sufficient detail to indicate compliance with applicable codes. Code review drawings are a key component in JOC's ability to accelerate pre-construction time lines. The contractor must ensure regulatory compliance within the contents of the drawings, which are submitted to the owner for approval as a component of the JO proposal.

Any project that impacts applicable life safety, building codes, or statutes, such as the ADA, should require, at a minimum, a field-dimensioned, proportionally scaled schematic drawing encompassing the limits of construction with compliancy details as needed. It is usually required that plans and drawings be developed electronically in CADD, compatible with the owner's current software applications. The owner can assist the contractor by furnishing any existing CADD files of as-built drawings, as appropriate, for the project during scoping.

### Special Project Procedures

If not addressed in other sections of the contract, this section provides instructions for contractors when special projects require environmental protection or handling of historical or archaeological finds during construction.

### Field Personnel

Typical contractor requirements for field staffing are outlined in this section. Included are any requirements for project management or supervision, dedicated or otherwise, or those associated with the level of worker activity in order to meet project completion times.

### Construction Progress Documentation

This section of the General Requirements addresses contractor requirements for status reports and scheduling of projects commonly found in other contracts, with some exceptions, as noted below.

### Status Reports

Status reports for JOC usually are required to be submitted monthly, and most identify and list the status of activity for all JOs that either have been requested or are in progress. Spreadsheets generally are designated, or a standard format that can be amended easily to include new JOs may be used.

### Scheduling

This section covers typical contractor requirements associated with the scheduling of work. Critical path methods, or variations thereof, for individual JOs usually are not required for projects of short duration (10–14 days). This is because the start and completion dates and performance times for all JOs are identified as components of the approved JO.

## Quality Requirements

This section of the General Requirements includes typical contractor requirements associated with quality control and inspections of work. A Quality Assurance/Quality Control Plan—issued within two weeks of contract award—addresses all aspects of quality control. It includes contact information for the contractor's quality control inspector, responsible for surveillance of work; documentation of unsatisfactory work; corrective actions; and interface with the owner's inspectors. Quality requirements for work performed within existing buildings should meet or exceed the existing quality of work-in-place. Materials should be new, unless otherwise specified, and all work should be performed in a professional manner by experienced workers, proficient in their trades, with trade credentials in accordance with any applicable standards, regulations, or statutes.

## Temporary Facilities & Controls

Numerous requirements associated with the contractor's office and/or field office, as well as mandates for the execution of a safe project, are covered in this section. Included is an example of a sequencing requirement allowing reasonable start-up time for the JOC contractor to mobilize operations and "gear up" to accommodate the owner's project volume. The remaining items in this section are to be addressed in the same manner as for other contract methods, with the exception of construction facilities.

---

**Temporary Facilities & Controls**

| | |
|---|---|
| Construction Start-up Period | Construction Facilities (cont.) |
| Temporary Utilities | Vehicular Access & Parking |
| Construction Facilities | Barriers & Enclosures |
| Temporary Construction | Project Signs |
| Construction Aids | Equipment Rental |

**Figure 5.7**

---

### Construction Start-up Period

Depending on the anticipated volume of work, it may be in the owner's best interest (as well as the contractor's) to include a requirement in the contract that allows for sequencing cumulative work volume over a designated start-up period. This will allow sufficient time after contract award for the contractor to have all personnel and equipment in place. Figure 5.8 depicts a sample construction start-up requirement to sequence the owner's cumulative work volume—anticipating $3 million worth of annual JOC work.

| Construction Start-up Period | |
| --- | --- |
| Calendar Days after Contract Implementation | Required Capability for Cumulative Value of Work |
| 21 Days | $0 |
| 30 Days | $300,000 |
| 60 Days | $1,000,000 |
| 90 Days | $3,000,000 |

**Figure 5.8** is an example showing milestone dates after contract award, allowing sufficient time for the contractor to put resources in place to sufficiently manage a cumulative value of work during the designated construction start-up period. The contractor may request work over specified capability requirements, but may decline work that exceeds the stated values without penalty if not yet sufficiently mobilized.

## Construction Facilities

Often, an area within the limits of the owner's facility (if applicable) is designated for the JOC contractor's field office and limited storage capability. Having a field office at the owner's facility improves contractor response times. While the owner provides the space for this office, the contractor bears the costs of any utility hookups and access costs. Other temporary facility and control requirements listed in Figure 5.7 can be developed to meet the owner's requirements.

## *Execution Requirements*

This section of the General Requirements includes standard requirements for cleaning and contract close-out. For JOC, contract close-out is referenced as job order close-out (since the contract is still in place). The only item unique to JOC from the standard list below is the requirement for the contractor to furnish record drawings or "as-built" drawings for each JO, with sufficient detail to document work performed in accordance with related requirements.

- Cleaning and rubbish handling
- Job order close-out
  - As-built drawings and testing reports
  - Operations and maintenance manuals & training
  - Warranties and guarantees
  - Certificates of occupancy, if applicable

The contractor's requirements for pre-final (punch list developed by owner's representative) and final inspection (all punch list items resolved by contractor) procedures, as well as warranty inspections, usually performed one year after project completion, are covered here.

## Facilities Operations

Facilities operations include commissioning, operations and maintenance, facilities maintenance, and moving equipment. If associated line items for these subdivisions are allowed and anticipated to be utilized during the term of the contract, appropriate requirements should be respectively included.

## Conclusion

The JOC General Requirements document is the primary component of the contract. The program start-up meeting sets the stage for open communication among key participants, and progress meetings are an effective way to maintain and develop mutually beneficial relationships by addressing issues and streamlining processes.

The General Requirements should identify and describe all the general execution requirements needed for successful contract management and project administration. All applicable processes and procedures should be addressed in detail. Each subdivision and section should clearly describe contractor requirements—reflecting the owner's needs and expectations and ensuring compliance with any applicable regulatory requirements during the term of the contract. Continual adherence to these requirements by the contractor, as well as consistent and fair contract management by the owner, plays an important role in the cost, time, and quality of each project authorized for accomplishment.

# Chapter 6
# Estimating JOC

Unit-price estimating takes more time to perform than less detailed types of estimating, such as square foot or assemblies. A higher level of accuracy is achieved when a project is properly unit priced, as is the case with JOC's job order pricing structure. There are many excellent references available to assist in estimating unit-price construction costs.

This chapter will provide repair and remodeling estimating tips and explain JO pricing as it relates to unit-pricing. It will also cover JOC coefficient development. Included are excerpts from RSMeans cost sources and tips for using minimum labor/equipment charges for small quantities.

## Training

Most JOC practitioners agree that comprehensive training by professionals in the discipline of unit-price estimating is important for estimators and reviewers. Owners' representatives who are unfamiliar with unit-price estimating should be trained before a JOC program is implemented. While experienced JOC contractors know how to unit-price estimate and may be proficient at identifying all allowable line items, training seminars can be useful as refreshers to fine-tune their skills. Joint training sessions with the owner's and contractor's staffs can ensure that all participants are using the proper procedures.

## Repair & Remodeling Estimating

Repair and remodeling presents the greatest challenge to estimators in the construction industry. Complete plans and specifications of the existing building are rarely available. The estimator may have only sketches by the owner to rely on or notes from the architect concerning the visible condition of the project. In some cases, openings must be cut through finish materials to examine the structure and/or infrastructure before planning modifications.

## Site Visit

After receiving alteration drawings and specifications, the first item on the agenda is to visit the site to scope the project. The estimator should inspect the entire interior and exterior conditions. Services to the building should be verified, as well as methods of access and egress, areas of storage, and local building regulations. The exterior inspection will provide clues to the type of construction, the age, and the general condition of the building.

The interior site visit should include a careful room-by-room inspection. It is helpful to outline exact alterations directly on the drawings for record purposes (before the takeoff is started). Room measurements should be verified and notes taken about the difficulty of material handling, limitations concerning storage of new materials, and methods of debris removal. Photos can be very helpful in this process.

Existing materials should also be noted, as the type of materials to be removed will affect costs. For example, when cutting openings to install new doors or windows, gypsum board partitions are much easier to work with than concrete masonry unit (CMU), plaster, or brick-faced walls.

## Cost Considerations

When a project requires removing walls, roofs, floors, or major structural items, it is imperative that the estimator collect as much knowledge as possible about the building. Although structural renovations usually are planned by an architect or engineer, the estimator must also be concerned with the cost of such items as temporary shoring, bracing, building protection, underpinning, and, in some cases, the protection of adjacent property.

When a new foundation will be added inside an existing building, consideration must be given to low overhead clearances for pile driving and the placement of concrete. For example, if steel piles are to be driven inside a basement, excess welding costs may be required for splicing piles. Another cost associated with structural modifications is steel cutting—which may also involve fire protection watches. Unless the steel is exposed and a fair estimate can be made of the linear feet of cutting, an allowance must be made to cover the cost of the unknown amounts of acetylene and oxygen and the labor-hours required to monitor activities of the steel cutter and carry out fire protection measures, if necessary.

One of the major judgments an estimator must make concerns potential delays caused by the facility's users who continue to occupy the affected space. The room-by-room visit should document areas where the facility's normal operations will continue, the extent and area of dust and noise protection required, and the location and size of existing equipment and furnishings that must be protected.

Anticipated delays in construction activities or utility shut-downs due to changes to existing utility service involving temporary loss of electrical, water, gas, or voice/data service must be documented and coordinated with the owner in advance. In some cases, the owner's operations may require evening or weekend work scheduling—and the corresponding wage-rate coefficients for non-standard hours might apply. A reduction in productivity must also be considered with the use of non-standard hours, since needed materials and labor resources are less available during the evening or early-morning hours.

Another factor beyond the estimator's control is the potential for unplanned work to emerge during the project. For example, if a ceiling is to be removed, it might be discovered that the floor joists are rotted and need replacement. In this case, the extra work should be brought to the owner's attention immediately and a change order initiated to cover the anticipated costs. The more knowledge about a building that an estimator has at the beginning, the lower the percentage of contingencies.

## JO Price Estimating

Accurate JO pricing is important to both the owner and the contactor. The JO price proposal is based on the project's scope of work and can be only as accurate as the project scoping is thorough—with or without design documents. UPB line items may not always fit exactly with the scope of work on a particular job. In these cases, reasonable selection of the closest-matching line item may be necessary. Unit-pricing should be quantified as accurately as possible and thoroughly checked after assembling the estimate. The primary requirements for JO pricing accuracy are as follows:

- The estimator's ability to accurately identify and quantify allowable UPB line items reflecting the agreed-on scope of work
- Proficient use of the designated UPB
- Correct application of the designated costs column and city cost index multiplier, as required
- Appropriate selection of the coefficient to be applied to the sum total of UPB line items in accordance with the required pricing structure
- Adherence to pricing requirements for non-prepriced items

### Minimum Labor/Equipment

The following excerpt from RSMeans *Facilities Construction Cost Data 2005*, Figure 6.1, explains how and when to adjust UPB pricing for minimum labor and equipment when dealing with small quantities. This is a common situation with JOC projects.

## Using Minimum Labor/Equipment Charges for Small Quantities

Estimating small construction or repair tasks often creates situations in which the quantity of work to be performed is very small. When this occurs, the labor and/or equipment costs to perform the work may be too low to allow for the crew to get to the job, receive instructions, find materials, get set up, perform the work, clean up, and get to the next job. In these situations, the estimator should compare the developed labor and/or equipment costs for performing the work (e.g., quantity x labor and/or equipment costs) with the *"minimum labor/equipment charge"* within that Unit Price section of the book.

If the labor and/or equipment costs developed by the estimator are LOWER THAN the *"minimum labor/equipment charge"* listed at the bottom of specific sections of Unit Price costs, the estimator should adjust the developed costs upward to the *"minimum labor/equipment charge."* The proper use of a *"minimum labor/equipment charge"* results in having enough money in the estimate to cover the contractor's higher cost of performing a very small amount of work during a partial workday.

A *"minimum labor/equipment charge"* should be used only when the task being estimated is the only task the crew will perform at the job site that day. If, however, the crew will be able to perform other tasks at the job site that day, the use of a *"minimum labor/equipment charge"* is not appropriate.

### 08500 | Windows

| 08550 | Wood Windows | | CREW | DAILY OUTPUT | LABOR-HOURS | UNIT | 2005 BARE COSTS | | | | TOTAL INCL O&P |
|---|---|---|---|---|---|---|---|---|---|---|---|
| | | | | | | | MAT. | LABOR | EQUIP. | TOTAL | |
| 200 0010 | CASEMENT WINDOW Including frame, screen, and grills | | | | | | | | | | 200 |
| 0100 | Avg. quality, bldrs. model, 2'-0" x 3'-0" H, dbl. insulated glass | | 1 Carp | 10 | .800 | Ea. | 176 | 27.50 | | 203.50 | 239 |
| 0150 | Low E glass | | | 10 | .800 | | 282 | 27.50 | | 309.50 | 355 |
| 0200 | 2'-0" x 4'-6" high, double insulated glass | | | 9 | .889 | | 228 | 30.50 | | 258.50 | 300 |
| 0250 | Low E glass | | | 9 | .889 | | 330 | 30.50 | | 360.50 | 410 |
| 0300 | 2'-3" x 6'-0" high, double insulated glass | | | 8 | 1 | | 395 | 34.50 | | 429.50 | 490 |
| 0350 | Low E glass | | | 8 | 1 | | 325 | 34.50 | | 359.50 | 415 |
| 8100 | Metal clad, deluxe, dbl. insul. glass, 2'-0" x 3'-0" high | | | 10 | .800 | | 175 | 27.50 | | 202.50 | 239 |
| 8120 | 2'-0" x 4'-0" high | | | 9 | .889 | | 211 | 30.50 | | 241.50 | 283 |
| 8140 | 2'-0" x 5'-0" high | | | 8 | 1 | | 240 | 34.50 | | 274.50 | 320 |
| 9000 | Minimum labor/equipment charge | | 1 Carp | 3 | 2.667 | Job | | 91.50 | | 91.50 | 151 |

## Example:

Establish the bid price to install two casement windows. Assume installation of 2' x 4' metal clad windows with insulating glass [Unit Price line number 08550-200-8120], and that this is the only task this crew will perform at the job site that day.

## Solution:

**Step One** — Develop the Bare Labor Cost for this task:

Bare Labor Cost = 2 windows @ $30.50/each = $61.00

**Step Two** — Evaluate the *"minimum labor/equipment charge"* for this Unit Price section against the developed Bare Labor Cost for this task:

*"minimum labor/equipment charge"* = $91.50 (compare with $61.00)

**Step Three** — Choose to adjust the developed labor cost upward to the *"minimum labor/ equipment charge."*

**Step Four** — Develop the bid price for this task (including O&P):

Add together the marked-up Bare Material Cost for this task and the marked-up *"minimum labor/equipment charge"* for this Unit Price section.

2 x ($211.00 + 10%) + ($91.50 + 65.7%)

= 2 x ($211.00 + $21.10) + ($91.50 + $60.12)

= 2 x ($232.10) + $151.62

= $464.20 + $151.62

= $615.82

**ANSWER:** $615.82 is the correct bid price to use. This sum takes into consideration the Material Cost (with 10% for profit) for these two windows, plus the *"minimum labor/equipment charge"* (with O&P included) for this section of the Unit Price book.

Figure 6.1

Once an estimate is complete, it should be reviewed—not only for accuracy—but for reasonable application. Understanding how to use a designated unit-price book, with all its divisions, subdivisions, indents, and reference number information, leads to proficient estimating. *(See "Unit-Pricing" in Chapter 2.)*

## Coefficient Development

Methodologies used for coefficient development vary slightly among JOC contractors, depending on contract requirements and applications, but common practice is to use known, actual costs from representative or "typical" past projects. These can be compared with a list of UPB line items that would be used for typical projects. *(See "Coefficients" in Chapter 2.)*

Contractors may or may not take into consideration the anticipated annual value of work to be performed, but most do. Contractors who have not worked for public-sector entities may have to adjust actual historical costs to compensate for increased requirements and conditions.

It is important that the coefficient is directly related to the designated UPB and overall pricing requirements of the JOC contract for which the coefficient will be utilized, as owner pricing requirements are not always the same. In general, coefficients developed and submitted to the owner for contracts that allow all or most of Division 1 line items will be lower than those submitted for contracts that do not allow, or greatly restrict, these line items in the JO pricing structure. This is because UPB line items for the general or prime contractor's O&P are located in Division 1. If they are not allowable as line items, contractors must include these costs within their coefficients to compensate their overhead costs and profit margins.

The UPB costs column and city cost multiplier designation have a significant impact on coefficient development. If the contract designates use of the "Bare Total" costs column, a contractor's coefficient will be higher than if the "Total, Incl. O&P" costs column is designated, since the latter column already includes the installing contractor's O&P.

Similarly, if city cost index multipliers are designated, coefficients will align to the localized area adjustment where work is to be performed, as opposed to alignment with national-average costs. This does not necessarily mean that projects will cost the owner more, or that contractors will be compensated less, with either costs column designation—or with or without city cost index multipliers. It means only that contractors will adjust coefficients during development to align with what is designated for use in contract requirements. *(See "City Cost Index Multipliers" in Chapter 5.)*

The following steps for coefficient development are based on the recommendations of The Cooperative Purchasing Network's (TCPN's) Area Job Order Contracting (AJOC©) program.

1. Identify the most recently completed small construction projects that represent a cross-section of trades. Ideally, they should be projects completed with the agency to which the contractor is bidding. If not, costs can be adjusted based on the differing terms, conditions, and so forth, as required.
2. For each project, gather the plans, specifications, and actual cost data.
3. Using the specified unit-price book, price the projects according to line items and quantities.
4. Divide the actual cost (adjusted for inflation or other factors, as required) by the total of the extended line items. Be sure to use the specified costs column and, if applicable, use the city cost index when computing the total price from the UPB. The result will normally be a decimal figure, such as .85 or 1.05 (price-book adjustment factor). Ideally, there will be several figures, each derived from the past projects being assessed, and the contractor can use a weighted average.
5. Based on the contractor's knowledge of the agency's budget and facilities, estimate in dollars the anticipated annual volume of work to be done.
6. Determine the overhead costs to manage the project at this volume. This will be the incremental cost of supervisory staff, vehicles, etc.
7. Divide this figure by the estimate of the actual cost of subcontracting or performing the work annually. This should be a decimal, such as .20.
8. Add the desired profit and headquarters overhead (G&A) percentages to the project overhead percentage.
9. Add 1.00 to this figure, and multiply it by the figure from step # 4. This decimal should be the bid coefficient for normal or standard hours. It can be adjusted based on the estimate of additional cost to perform the work during non-standard hours (e.g., nights and weekends).
10. Review the projects originally used. Multiply the total price for each project, as determined from the UPB, by the coefficient. The resulting total price for the project can be adjusted as necessary.

**Example:**

> Adjusted cost of project = $25,000.00
> Extended line items = $30,000.00; 25/30 = .83 (PBA)
> Estimated owner funding = $2.5M/year.
> Estimated cost of actual work = $2M/year
> Estimated overhead cost = $300K.
> Project overhead percentage = $300K/$2,000K = .15
> Profit = 8%
> Corp. OH = 2%
> Total = .25, or 1.25
> 1.25 × .83 = 1.038 as a "normal" coefficient. Adjust for "other-than-normal."

The above example is simplified for the purpose of illustration only. Contractors who are inexperienced in JOC projects may choose to start by subcontracting with an established JOC contractor before taking on their own JOC contracts.

## Conclusion

Proficiency at unit-price estimating is essential for successful JOC contracting, since compensation to the contractor for work performed is a lump-sum fixed price based primarily on the estimated cost of the project as derived from the scope of work. Cost estimating accuracy begins with reliable, comprehensive, up-to-date, easily understood cost data, such as that developed by RSMeans.

Proper training for estimators and reviewers alike on the use of a JOC contract's designated UPB in accordance with pricing structure requirements will help ensure accurate estimates derived through tried and proven procedures. In addition, a clear understanding of the owner's JO pricing structure requirements plays an important role in coefficient development.

Estimating is a straightforward process using established procedures based on information derived from new design documents, as-builts, project site visits, and a thorough scoping process. Even so, 100% objectivity in cost estimating is elusive for even the most adept estimators, and a higher measure of judgment is needed when estimating repair and remodeling projects as compared with new construction.

# Chapter 7

## Anatomy of a Job Order

This chapter explains the JO authorization process and presents standardized documents for a typical project. Generally, a *job order authorization* form documents the following:

- The JO proposal request to the contractor, including any attached documents
- The JO proposal from the contractor, including the scope of work
- The JO price proposal, including UPB breakdown
- The written authorization by the owner to accomplish the described work

## Request for a Job Order Proposal

The first step in the process of JO authorization is a written request from the owner to the contractor for a JO proposal for the work described. If available, reference documents and/or record drawings may be attached to the request, as deemed appropriate by the owner. If the owner has record drawings in CADD format and is proficient in the use of designated program management interface tools, this initial procedure can be accomplished within minutes. Sometimes a simple photo may accompany the JOC proposal request.

Figure 7.1 is a sample JO authorization form. The drawings that follow in Figure 7.2 are sample owner's record drawings (as-builts) of the building where the work is to take place. *(Note that these correspond to the "Drawings" check box in the first section of the form as an "Attachment" item.)* Together these documents are transmitted to the contractor at this initial phase of the process to assist with scoping and subsequently allow a base format for the contractor to build on that shows the owner's desired modifications. All of these documents can be emailed easily as attachments among project participants.

## Request for Job Order Proposal

Facilities Owner Logo with Hyperlink (optional)

**XYZ Facilities Owner**
Street
City, State  Zip
Phone:                                    FAX:
Web site: (Hyperlink)

| JOB ORDER AUTHORIZATION | | |
|---|---|---|
| Contract Proposal #: XYZ-JOC-1 | JO #: 00-1234-00 | P.O.#: |
| Job Order Title: Main Office Building, 3rd Floor  Remodel | Location: Main Office Building | |

### REQUEST FOR JOB ORDER PROPOSAL

To:      JOC Contractor                                                          From: XYZ Facilities Owner

Please provide a Proposal for the above named Job Order as described below:

Description of Work: Remodel the 3rd floor of the Main Office Building, converting existing Employee Lounge and Kitchen areas into a Conference Room and Office. Use PWR coefficient and our Standards of Construction as applicable.

This Job Order Requires:
- ☐ Testing        ☐ A/E Design        ☒ Submittals        ☒ Standard Hours
- ☐ Emergency Mobilization    ☐ Phasing    ☒ Code Review Drawings    ☒ Permitting    ☐ Non-Standard Hours
- ☐ Budget Assistance    ☒ Shop Drawings    ☒ Sequence Schedule

Links:       http://www.rsmeans.com/

| JO Start Date | ☐ Required | JO Completion Date | ☐ Required | Architect/Engineer |
|---|---|---|---|---|
| 11/1/YR | ☒ Desired | 12/15/YR | ☒ Desired | |

To schedule a site visit, contact:
Name:     (drop down box with Owner's Rep. Names)
Phone:    (phone #'s)        Cell:    (cell #'s)
Email:    (drop down box with Owner's Rep. Email

Attachments:
- ☐ None        ☒ Drawings        ☐ Other (List Below)
- ☐ Specifications    ☐ Samples    1.
                                   2.

XYZ Representative _____    10/15/YR
Facilities Contract Manager or Administrator        Date

### JOB ORDER PROPOSAL

To: XYZ Facilities Owner                    From:

Contact: _____    Phone: _____
Street/P.O. Box: _____
City: _____    State: _____    Zip: _____
Cell: _____    Fax: _____
Email: _____

Attachments:
- ☐ Scope of Work
- ☐ Price Proposal with UPB Breakdown
- ☐ Estimated Start & Completion Dates

Our Price Proposal, inclusive of the above-identified attachments, is hereby submitted for your consideration. All work is to be in accordance with the Contract Proposal identified above, inclusive of related documents contained or referenced therein.

Contractor's Authorized Representative _____    Date

### JOB ORDER AUTHORIZATION

To:                                                          From: XYZ Facilities Owner

The Job Order described above is hereby authorized for accomplishment, contingent upon the above-identified open Purchase Order.

1. Amount not to exceed _____

   Dollars $ _____ .

2. The above amount includes a Base Proposal amount of $ _____ Dollars and an Owner Contingency

   amount of $ _____ Dollars.

3. Completion date is on or before _____ (days, date).

4. Liquidated damages in the amount of $ _____ per calendar (day) (week) shall accrue to the Owner for late completion of this Authorization.

5. Prior to beginning the work, provide a HUB Plan (if applicable) and ensure execution of Payment and Performance Bonds as required in accordance with the Provisions of the Contract Proposal. Submit file-stamped copies to the Owner for the record.

Owner-authorized name _____    Signature _____    Date

Owner-authorized name _____    Signature _____    Date

**Figure 7.1** is a sample JO authorization form, used to document the JO authorization process. The owner would fill out all JO-related identifiers in the top section and all information in the request section of the form. The form, along with any attachments, is then sent to the contractor.  In this case, the owner is attaching applicable record drawings of the building to the request.[1]

## Building Record Drawings

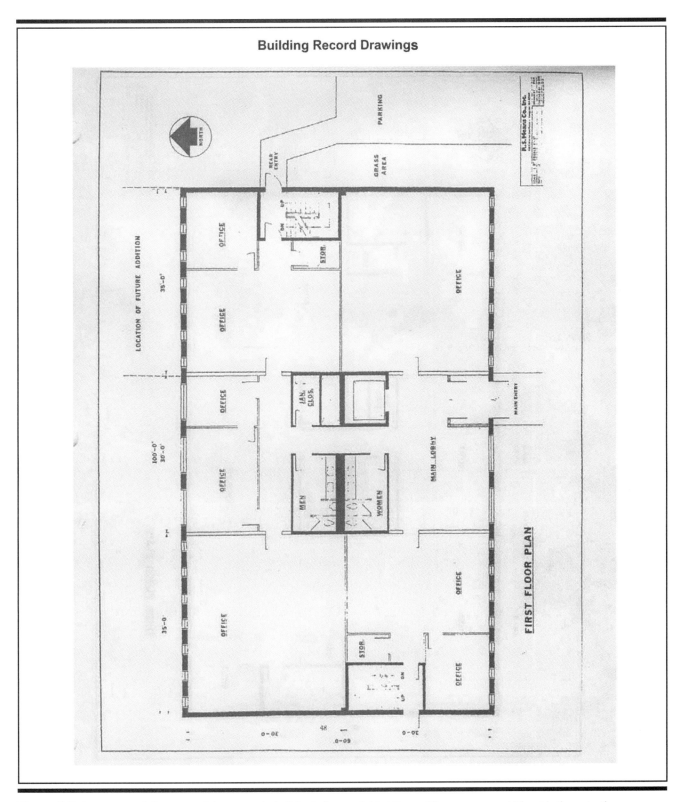

**Figure 7.2** shows record drawings of the example building's first and third floors. The owner would furnish these to the contractor to assist in scoping the project. Although the project is located on the third floor, the first floor plan helps with planning material handling, staging, and access.

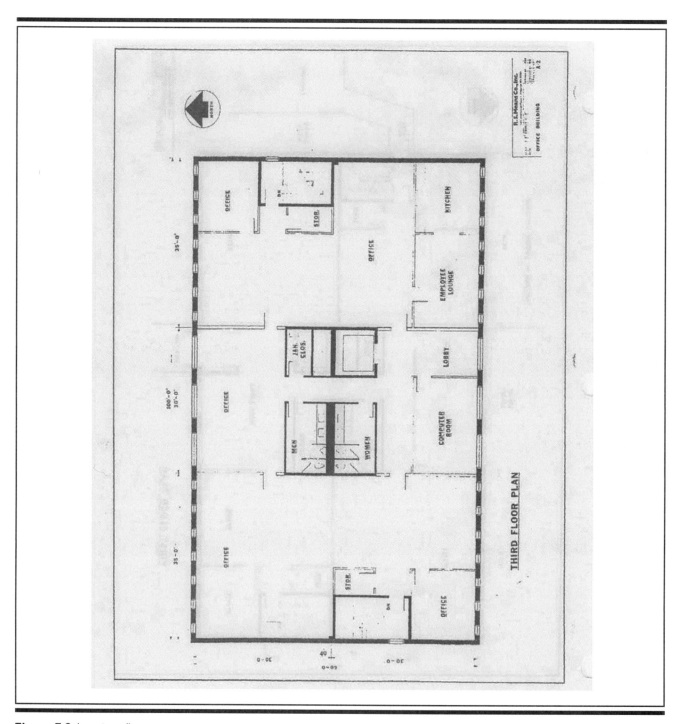

**Figure 7.2** (continued)

## Job Order Proposal Submittal

After the owner requests a JO proposal, the contractor responds by contacting the identified owner's representative within the required response time. The site visit can then be scheduled and the project scoped. *(See "Job Order Authorization" in Chapter 5.)* The contractor returns the form to the owner, documenting the JO proposal submittal with attached documents as required (scope of work and JO price proposal).

## Job Order Proposal Submittal

**XYZ Facilities Owner**
Street
City, State Zip
Phone:                FAX:
Web Site: (Hyperlink)

### JOB ORDER AUTHORIZATION

| Contract Proposal #: XYZ-JOC-1 | JO #: 00-1234-00 | P.O.#: |
|---|---|---|

| Job Order Title: Main Office Building, 3rd Floor Remodel | Location: Main Office Building |
|---|---|

### REQUEST FOR JOB ORDER PROPOSAL

To:      JOC Contractor                                      From: XYZ Facilities Owner

Please provide a Proposal for the above named Job Order as described below:

Description of Work: Remodel the 3rd floor of the Main Office Building, converting existing Employee Lounge and Kitchen areas into a Conference Room and Office. Use our Standards of Construction, as applicable.

This Job Order Requires:
- ☐ Testing
- ☐ A/E Design
- ☒ Submittals
- ☒ Standard Hours
- ☐ Emergency Mobilization
- ☐ Phasing
- ☒ Code Review Drawings
- ☐ Non-Standard Hours
- ☐ Budget Assistance
- ☒ Shop Drawings
- ☒ Sequence Schedule
- ☒ Permitting

Links:      http://www.rsmeans.com/

| JO Start Date | ☐ Required | JO Completion Date | ☐ Required | Architect/Engineer |
|---|---|---|---|---|
| 11/1/YR | ☒ Desired | 12/15/YR | ☒ Desired | |

To schedule a site visit, contact:
Name:      (drop down box with Owner's Rep. Names)
Phone:      (phone #'s)     Cell:     (cell #'s)
Email:      (drop down box with Owner's Rep. Email

Attachments:
- ☐ None
- ☒ Drawings
- ☐ Other (List Below)
- ☐ Specifications
- ☐ Samples

1. _____
2. _____

XYZ Representative _____ 10/15/YR
Facilities Contract Manager or Administrator          Date

### JOB ORDER PROPOSAL

To: XYZ Facilities Owner                    From:     JOC Contractor

Contact:      Contractor Representative      Phone:      000-123-0000
Street/P.O. Box:      Any Street
City     Any City                    State:     Any     Zip:     00000-0000
Cell:     000-123-4567                    Fax:     000-123-4567
Email:      JOC Contractor Representative@Email address

Attachments:
- ☒ Scope of Work
- ☒ Price Proposal with UPB Breakdown
- ☒ Estimated Start & Completion Dates

Our Price Proposal, inclusive of the above-identified attachments, is hereby submitted for your consideration. All work is to be in accordance with the Contract Proposal identified above, inclusive of related documents contained or referenced therein.

JOC Contractor _____ 10/23/YR
Contractor's Authorized Representative          Date

### JOB ORDER AUTHORIZATION

To:                                      From: XYZ Facilities Owner

The Job Order described above is hereby authorized for accomplishment, contingent upon the above-identified open Purchase Order.

1. Amount not to exceed _____

   Dollars $ _____ .

2. The above amount includes a Base Proposal amount of $ _____ Dollars and an Owner Contingency

   amount of $ _____ Dollars.

3. Completion date is on or before _____ (days, date).

4. Liquidated damages in the amount of $ _____ per calendar (day) (week) shall accrue to the Owner for late completion of this Authorization.

5. Prior to beginning the work, provide a HUB Plan (if applicable) and ensure execution of Payment and Performance Bonds as required in accordance with the Provisions of the Contract Proposal. Submit file-stamped copies to the Owner for the record.

| Owner-authorized name | Signature | Date |
|---|---|---|
| Owner-authorized name | Signature | Date |

**Figure 7.3** shows the contents of the form when it is completed by the contractor with attached documents noted. This section of the form is completed after the scope of work has been agreed on and the JO price proposal has been approved.

## Scope of Work Submittal

The following is the agreed-on scope of work submitted for the example project from Figures 7.1–7.2. Using the MasterFormat classification system helps standardize documents. This provides consistency and familiarity throughout the process among key participants in the exchange of documents.

---

### SCOPE OF WORK

**From:** JOC Contractor         **Date:** 10/20/YR

**To:** XYZ Facility Owner

**Project Name:** Main Office Building, 3rd Floor Remodel

**Project Location:** Main Office Building, 3rd Floor, 123 Any Street, City, State

**Project Room Number or Area:** Kitchen and Employee Lounge

**Scope Developed by:** (Contractor's Representative)    **Site Visit Date:** 10/17/YR

**Start Date:** On or before 11/1/YR      **Completion Date:** On or before 12/15/YR

#### Summary of Scope of Work:

1. The work to be done is located on the third floor of the building, accessible by the driveway and parking area southeast of the building.
2. The small passenger elevator off the main lobby may not be used to transport materials. The stairwell at the east side of the building near the work area is to be used for workers' access to the job site and all deliveries.
3. The work area must be completely closed off to keep the remainder of the building absolutely dust-free to protect computer equipment.
4. The building will be occupied from 7:00 a.m. to 6:00 p.m. on weekdays. All reasonable precautions must be taken to prevent or minimize disruption of user operations.
5. The existing kitchen and employee lounge are to be renovated to accommodate changes in use. The existing plumbing lines are to be capped off at the floor level for future use. The existing kitchen vent duct will be capped off above the new ceiling assembly. One new exterior window will be installed to match the existing window.
6. Freezing temperatures are a possibility during the work schedule. All masonry work must be protected from freezing.

#### 1. General Requirements:

Scaffolding will be required to access the new window area. Tarpaulins will be needed to protect the openings and new brickwork from the elements. All areas of work and access routes must be cleaned prior to final inspection. A draft sequence schedule is attached with this proposal, with a final version to be submitted after the JO authorization and prior to mobilization at the project site. The JOC contractor will obtain the required building permit for this project.

#### 2. Site Construction:

Work includes select interior demolition of existing non-load-bearing partition walls, ceiling assembly, etc., as needed, and saw-cutting of masonry at the interior wall for a new opening from the existing lobby and at the exterior wall

---

for a new window opening. A rubbish chute from the third floor to the dumpster will be located below the new exterior opening for the duration of the project.

## 3. Concrete:

Not applicable.

## 4. Masonry:

Work includes selective demolition of masonry units and installation of pre-cast concrete lintels, CMUs, and standard common bricks. Cut and patch to match existing windows.

## 5. Metals:

Work includes installation of steel angle lintels at the new interior opening and at the new window opening.

## 6. Wood & Plastics:

Work includes installation of an oak sill at the interior of the new window to match existing windows.

## 7. Thermal & Moisture Protection:

Work includes installation of ceiling and wall insulation, flashing at the new window opening, and caulking, as needed, to seal the new window.

## 8. Doors & Windows:

Work includes installation of steel door frames, hand-carved mahogany wood doors, door hardware and locksets, and an exterior window with an aluminum frame. All are to match the existing doors and windows. The lead time for doors is four to five weeks from the date of order.

## 9. Finishes:

Work includes installation of metal stud partition wall assemblies with 5/8" gypsum board, acoustical ceiling assemblies, carpet, and cove base.

## 10. Specialties:

Not applicable.

## 11. Equipment:

Not applicable.

## 12. Furnishings:

Not applicable.

## 13. Special Construction:

Not applicable.

## 14. Conveying Systems:

Not applicable.

## 15. Mechanical:

Work includes capping the existing plumbing lines below floor level.

## 16. Electrical:

Materials include 3 light switches, 16 duplex receptacle outlets, 4 voice/data boxes with 3/4" emt extended to 6" above the ceiling assembly, and 18 ea. fluorescent light fixtures (match existing) in locations to be designated by the owner and documented in record drawings. Existing branch circuitry will be reconfigured as needed; additional electrical service is not required. Voice/data box covers, wiring, and terminations to existing devices are through a separate contract with the owner. Any utility outages will be brief and scheduled in advance with the owner's project representative.

### Shop Drawings

The following shop drawings will be submitted for the materials specified (or identified to be furnished) for this project. Shop drawings for common items identified within the Standards of Construction (previously approved and on file for the owner's record) will not be re-submitted unless requested, or unless there have been significant changes since the last approval.

- ☐ Reinforcing Steel, Structural Steel
- ☒ Architectural Pre-Cast Concrete
- ☒ Stonework or Masonry
- ☐ Miscellaneous Metal
- ☐ Millwork and Casework
- ☒ Metal Doors, Frames, Windows, and Glazing
- ☒ Wood Doors
- ☐ Aluminum Storefront and Curtain Wall
- ☒ Finish Hardware and Accessories
- ☐ Toilet Partitions
- ☐ Special Equipment
- ☐ Waterproofing
- ☒ Finish Materials
- ☐ Elevators
- ☒ Painting
- ☐ Mechanical Equipment and Controls
- ☒ Electrical Fixtures and Equipment

### Samples

Samples of the following materials will be submitted: carpet, cove base, and exterior brick.

### Mockups and Field Samples

Field samples will be required for wall finishes and brick work.

### Permits

This project will require a city building permit.

### Code Review Drawing

Incidental code review drawings showing the proposed interior modifications with the exterior wall detail at the new window opening are included with this proposal. (See attached.) Record drawings will be furnished at the project closeout.

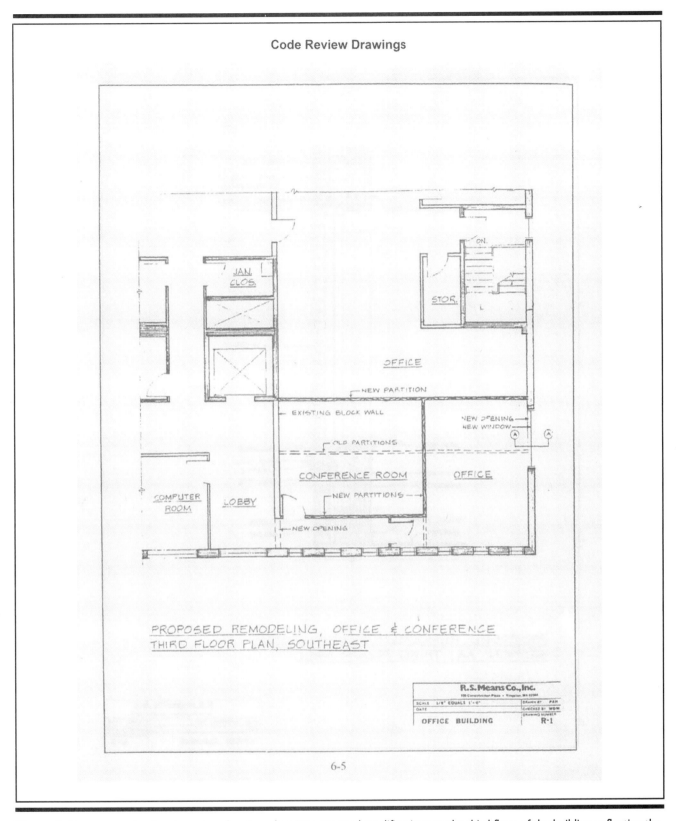

PROPOSED REMODELING, OFFICE & CONFERENCE
THIRD FLOOR PLAN, SOUTHEAST

6-5

**Figure 7.4** is an example of code review drawings showing proposed modifications to the third floor of the building, reflecting the owner's intent as discussed during the site visit. (These drawings should be attached to the Scope of Work.) The contractor can reduce time spent developing drawings with the use of CADD software compatible with the owner's software, and by referencing specific standard design details found in reference books such as *Architectural Graphic Standards*, published by John Wiley & Sons, Inc. Any designated additions, deletions, or changes to those details should be noted (provided this method is approved by the owner).

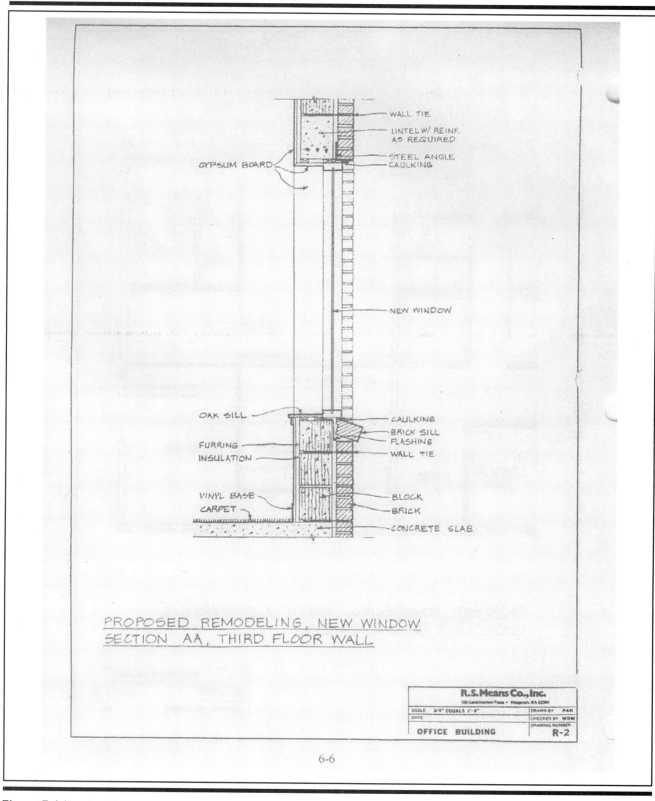

WALL TIE
LINTEL W/ REINF. AS REQUIRED
STEEL ANGLE
CAULKING

GYPSUM BOARD

NEW WINDOW

OAK SILL
CAULKING
BRICK SILL
FLASHING
FURRING
WALL TIE
INSULATION

VINYL BASE
BLOCK
CARPET
BRICK
CONCRETE SLAB

PROPOSED REMODELING, NEW WINDOW
SECTION AA, THIRD FLOOR WALL

R.S. Means Co., Inc.
100 Construction Plaza • Kingston, MA 02364

| SCALE | 3/4" EQUALS 1'-0" | | DRAWN BY | PAM |
| DATE | | | CHECKED BY | WDM |
| OFFICE BUILDING | | | DRAWING NUMBER | R-2 |

6-6

**Figure 7.4** (continued)

# JO Price Proposal Submittal

Once the owner approves the scope of work, the contractor submits the JO price proposal, reflecting the project's scope, to the owner for review. The following is a sample JO price proposal. (All costs are for illustrative purposes only and are not intended to reflect actual project costs.)

---

**JOB ORDER PRICE PROPOSAL**

**From:** <u>JOC Contractor</u>                                        **Date:** 10/22/YR

**To:** <u>XYZ Facility Owner</u>

**Project Name:** Main Office Building, 3<sup>rd</sup> Floor Remodel

**Project Location:** Main Office Building, 3<sup>rd</sup> Floor, 123 Any Street, City, State

**Project Room Number or Area:** Kitchen and Employee Lounge

**Price Proposal Developed by:** (<u>Contractor's Representative</u>)

---

## Summary of Price Proposal

| Divisions | Total, Incl. O&P | Notes |
|---|---|---|
| Division 1 General Requirements | $ 1,109.20 | |
| Division 2 Site Construction | 12,797.00 | |
| Division 3 Concrete | 0.00 | |
| Division 4 Masonry | 3,033.44 | |
| Division 5 Metals | 392.07 | |
| Division 6 Wood and Plastics | 49.00 | |
| Division 7 Thermal and Moisture Protection | 1,129.92 | |
| Division 8 Doors and Windows | 5,252.98 | |
| Division 9 Finishes | 19,109.56 | |
| Division 10 Specialties | 0.00 | |
| Division 11 Equipment | 0.00 | |
| Division 12 Furnishings | 0.00 | |
| Division 13 Special Construction | 0.00 | |
| Division 14 Conveying Systems | 0.00 | |
| Division 15 Mechanical | 72.50 | |
| Division 16 Electrical | 6,585.00 | |
| Subtotal of Divisions | 49,431.00 | (Rounded to nearest dollar) |
| City Cost Index @ 1.00 | 49,431.00 | |
| Coefficient @1.00 | 49,431.00 | |
| Subtotal | 49,431.00 | |
| Contingency @ 8% | 3,954.00 | |
| **Total** | **$53,385.00** | |

**Figure 7.5** is a summary of the price proposal organized by CSI MasterFormat divisions. In this case, the "Total Incl. O&P" costs column has been designated for use in the contract. The applicable coefficient is used to modify the subtotal of divisional pricing. Here, the applicable coefficient is 1.00, so no modification is necessary. The city cost index multiplier shown in this example identifies national average UPB unit-cost data—contract requirements may vary. *(See "City Cost Index Multipliers" in Chapter 5.)* The UPB line item breakdown (not shown) should accompany the JO price proposal summary as an attachment.

### UPB Breakdown

In addition to the summary shown in Figure 7.5, a complete detailed listing of UPB line items with associated quantities is submitted by the contractor as an attachment to the scope of work summary. The purpose of this listing, commonly referred to by JOC practitioners as the "UPB breakdown," is to identify the desired scope of work. *(See "Unit-Pricing," in Chapter 2 for an example of the UPB breakdown.)* The JO's price proposal UPB breakdown should comply with the contract's JO pricing requirements.

Estimators' techniques to build unit-price estimates vary, but most use the "build it before you build it" method. In doing so, the line items sequentially entered into an estimate's cost list may not be categorically grouped in divisional format. However, most cost-estimating software applications have the capability to sort in numerical order during or after the process of entering line items. This function will categorize line items by division. Once the price proposal is complete, it is reviewed for accuracy by the owner and any adjustments are made before approval. Once approved, the total price and contingency amounts (if necessary) are transferred to the final section of the authorization form.

## Job Order Authorization

After the owner has approved the scope of work (with performance time lines) and the price proposal, the JO can be authorized for accomplishment, as shown in Figure 7.6, a completed job order authorization form.

## Job Order Authorization

Facilities Owner Logo with Hyperlink (optional)

**XYZ Facilities Owner**
Street
City, State  Zip
Phone:                    FAX:
Web site: (Hyperlink)

| JOB ORDER AUTHORIZATION | | |
|---|---|---|
| Contract Proposal #: XYZ-JOC-1 | JO #: 00-1234-00 | P.O.#:  00-URAGO-00 |
| Job Order Title: Main Office Building, 3rd Floor  Remodel | | Location: Main Office Building |

### REQUEST FOR JOB ORDER PROPOSAL

To:     JOC Contractor

From: XYZ Facilities Owner

Please provide a Proposal for the above named Job Order as described below:

Description of Work: Remodel the 3rd floor of the Main Office Building, converting existing Employee Lounge and Kitchen areas into a Conference Room and Office. Use our Standards of Construction, as applicable

This Job Order Requires:
☐ Testing   ☐ A/E Design   ☒ Submittals   ☒ Standard Hours
☐ Emergency Mobilization   ☐ Phasing   ☒ Code Review Drawings   ☒ Permitting   ☐ Non-Standard Hours
☐ Budget Assistance   ☒ Shop Drawings   ☒ Sequence Schedule

Links:     http://www.rsmeans.com/

| JO Start Date | ☐ Required | JO Completion Date | ☐ Required | Architect/Engineer |
|---|---|---|---|---|
| 11/1/YR | ☒ Desired | 12/15/YR | ☒ Desired | |

To schedule a site visit, contact:
Name:     (drop down box with Owner's Rep. Names)
Phone:    (phone #'s)     Cell:     (cell #'s)
Email:    (drop down box with Owner's Rep. Email)

Attachments:
☐ None        ☒ Drawings        ☐ Other (List Below)
☐ Specifications   ☐ Samples          1.
                                       2.

XYZ Representative                                10/15/YR
Facilities Contract Manager or Administrator         Date

### JOB ORDER PROPOSAL

To: XYZ Facilities Owner          From:     JOC Contractor

| Contact:     Contractor Representative | Phone:     000-123-0000 | Attachments: |
|---|---|---|
| Street/P.O. Box:     Any Street | | ☒ Scope of Work |
| City     Any City | State:   Any     Zip:     00000-0000 | ☒ Price Proposal with UPB Breakdown |
| Cell:   000-123-4567 | Fax:     000-123-4567 | ☒ Estimated Start & Completion Dates |
| Email:     JOC Contractor Representative@Email address | | |

Our Price Proposal, inclusive of the above-identified attachments, is hereby submitted for your consideration. All work is to be in accordance with the Contract Proposal identified above, inclusive of related documents contained or referenced therein.

JOC Contractor                                10/23/YR
Contractor's Authorized Representative          Date

### JOB ORDER AUTHORIZATION

To:     JOC Contractor                          From: XYZ Facilities Owner

The Job Order described above is hereby authorized for accomplishment, contingent upon the above-identified open Purchase Order.

1. Amount not to exceed     Fifty-Three Thousand, Three Hundred Eighty-Five

   Dollars $     53,385.00              .

2. The above amount includes a Base Proposal amount of $     49,431.00          Dollars and an Owner Contingency

   amount of $     3,954.00     Dollars.

3. Completion date is on or before     12/15/YR              (days, date).

4. Liquidated damages in the amount of $ _____ per calendar (day) (week) shall accrue to the Owner for late completion of this Authorization.

5. Prior to beginning the work, provide a HUB Plan (if applicable) and ensure execution of Payment and Performance Bonds as required in accordance with the Provisions of the Contract Proposal. Submit file-stamped copies to the Owner for the record.

| (Owner-authorized name) | (Owner signature #1 as appropriate) | 10/25/YR |
|---|---|---|
| Owner-authorized name | Signature | Date |
| (Owner-authorized name) | (Owner signature #2 as appropriate) | 10/25/YR |
| Owner-authorized name | Signature | Date |

**Figure 7.6** is a completed JO authorization form as would be issued to the contractor.  It documents the owner's approval of the contractor's JO proposal to accomplish the JO, contingent upon the issuance of a purchase order (if required) and any documents referenced in Section #5 of the form.

## Conclusion

This chapter illustrates sequenced procedures for JO authorization in accordance with JOC General Requirements, showing direct application of related contract provisions to a typical JOCable renovation project. The figures show project-related design documents depicting the level of design effort needed to accomplish work, and standardized forms developed exclusively for JOC contract execution. When possible, these documents are in alignment with the MasterFormat system of classification.

These documents and forms contain examples of specific project information that can be rapidly exchanged among the owner, contractor, and design professional (as needed) via designated program-management interface tools. The documents are the primary components needed to capture the owner's project concept or intent and document the JO authorization process. The sequential steps shown in this representative project authorization can be used as a guide for all projects to be administered through a JOC contract.

The use of standardized forms in JO authorization provides structure, consistency, and auditable documentation to the JO authorization process. With practice in the use of forms like these, program participants become familiar with, and begin to appreciate, the expeditious nature of the pre-construction aspects of JOC. Using standardized forms and procedures also helps the contractor meet contract requirements and lays the groundwork for key program participants to collaborate in streamlining processes.

---

1. The Job Order Authorization form shown in Figure 7.1 is a modified version of the form originally developed and successfully implemented by Joe Martin, PE, JOC pioneer of program implementation for large public educational facilities. Mr. Martin is currently director of engineering services for Northside Independent School District in San Antonio, Texas, where he is responsible for providing oversight of all district plans and construction, including NISD's JOC program. Modifications to the contents of the request section have been made to include additional project-specific requirements and attachments (formatting by Brenda Henderson), with input from the facilities planning, design, and construction contract administration staff at Texas State University-San Marcos, as well as several JOC contractors.

# Chapter 8

# The Partnering Aspect of JOC

The principle of partnering is interwoven with a successful JOC program. This chapter will address current thought on partnering and its attributes in general application. A brief history of partnering is included as background, followed by examples of partnering benefits that can be achieved with JOC. Explanations are offered about how, when, and why JOC methodologies and strategic-partnering principles connect.

Not all individuals are good candidates for interfacing with partnering principles, because some don't possess the attributes needed to successfully engage in these arrangements. This chapter will explore some of these characteristics, since they are crucial for successful JOC projects.

## Understanding Partnering Principles

Several noted experts on partnering, as well as many professional organizations, offer varying definitions of partnering as it pertains to the construction industry. However, there is a general consensus that partnering is more than individuals working in teams to achieve mutually defined goals.

Partnering, which stresses "win-win" relationships through openness and trust, is best characterized as a set of collaborative processes. These processes emphasize the importance of common goals and raise questions as to how goals are agreed on, at what level they are specified, and how they are articulated.[1] One might also ask how the processes are monitored for effectiveness and how they are modified, if necessary, to stay on course.

Most literature reflects the generic definition of partnering to include the following:

- A set of collaborative processes (as opposed to a relationship only)
- A commitment to achieve mutually identified objectives between two (or among more) parties or organizations based on cooperation, open communication, and heightened effectiveness through continuous improvement.

The following is an outline of shared values derived from an excerpt of a U.S. Department of Energy partnering agreement offered as an example of strategic partnering.

*Teamwork* – Partners each take an active role in achieving the mission and resolving problems or challenges, with the collective and cohesive team being more powerful than either of its parts.

*Mutual Trust* – Partners' actions are consistent and predictable. Trust is earned when actions are proven to be consistent with words and commitments.

*Respect* – Partners each understand and honor the other's authorities, boundaries, and honest differences.

*Open Communication* – Partners encourage communication and sharing at all levels.

*Honesty* – Partners have the courage to confront and resolve conflict in an open and forthright manner.

*Accountability* – Partners recognize and accept commitments and choices, and are responsible to each other for meeting commitments.

*Continuous Improvement* – Partners recognize the value of change in the ways of doing business, which eliminates unnecessary processes and procedures that provide little benefit to either party.

## Partnering in Construction

Partnering in the construction industry can be categorized as "project-partnering" when applied to individual projects (usually large in scope and scale) or as "strategic partnering" involving long-term collaborative processes (therefore applicable to JOC). According to the Construction Industry Institute Australia (CIIA)[2], typical benefits from partnering are as follows:

- Reduced exposure to litigation
- Improved project outcomes in terms of cost, time, and quality
- Lower administrative and legal costs
- Increased opportunity for innovation and value engineering
- Increased chances for financial success

There is a long history of partnering in contractual arrangements. This is especially true in the manufacturing industry, where collaborative processes are easily developed due to the partners' similarities in operations and goals. The construction industry faces a more diverse set of participants, with little or no operational similarities and more diverse goals. This makes partnering more challenging.

The construction industry has long practiced teamwork, which has proved to be a necessary component to successful projects. Can people work together in teams, ultimately achieving their goals, without getting along with each other? Absolutely. Does the construction industry have a long history of disputes and conflicts as a result of confrontational, adversarial relationships? Certainly. There is no question that these types of relationships can hinder efficient and effective performance and cause distractions from key issues and goals. Disputes may result in time- and cost-consuming mediation, arbitration, and, sometimes, litigation.

Furthermore, some observers suggest that the traditional public procurement requirements of compulsory competitive bidding (based primarily on low-bid contract award) are detrimental to collaborative processing in the industry. The implication is that these requirements promote adversity, since awardees seek ways to cut costs to make up for lost profit margins (sacrificed to beat the competition). A lesser-quality performance by the contractor is coupled with a propensity to award subcontractors and material vendors based primarily on pricing. There is less emphasis overall on quality and performance.

Consequences have been disputes, delays, and increased contract management efforts by the owner and/or A/E in order to curb poor performance and challenge inflated change orders. Litigation has increased in the industry. As recently as the 1990s, many contractors were faced with escalating legal fees, which cut further into their profit margins. Meanwhile, owners were doing everything they could to transfer or mitigate risks—including, in part, more stringent contract provisions with emphasis on past performance, quality, and management/technical abilities.

Partnering is a glimmer of hope for an industry with a history wracked by divergent interests and mistrust. Partnering principles are gaining momentum—as a welcomed trend, and as a new beginning to a dynamic industry. Formal partnering agreements in the construction industry are gaining in popularity, being endorsed by organizations such as the U.S. Department of Energy (DOE), the U.S. Army Corps of Engineers (USACE), the Associated General Contractors of America (AGC), the Construction Industry Institute (CII), and many others, both in the United States and internationally.

## JOC Partnering

Construction does not always align with strategic partnering, especially in the case of small, short-term projects with traditional procurement regulations regarding contract award. (USACE defines "small projects" as those costing $1M or less.) While these delivery methods can benefit from project partnering, all collaborative processing tends to end before the benefits of long-term relationships can develop. JOC is at the forefront of replacing traditionally adverse relationships with mutually beneficial relationships using strategic-partnering.

Partnering with JOC includes partners learning each other's processes as soon as possible after contract award. This involves identifying shared values, openly addressing apprehensions, and collaborating to define mutual goals. This is also the time to identify potential problems and generate solutions. Setting performance standards, monitoring actions, evaluating goal commitments, and acknowledging achievements are equally important; otherwise the partnering effort may be lost. JOC partnering is based on mutual trust, which can be earned only through daily demonstrations of competence, character, and professionalism. *(See "Program Start-up Meeting" and "Program Progress Meetings" in Chapter 5.)*

## Benefits of JOC Partnering

The fact is, unless the contractual relationship is terminated, owners and JOC contractors are "joined at the hip" for the duration of the contract term. They can help each other achieve success with the following actions:

- Ensure lines of communication are kept open with clear understanding of individual roles and chains of authority.
- Meet schedule and budget targets.
- Provide users with safe and functional facilities.
- Streamline processes and reduce cycle times.
- Resolve disputes immediately.
- Minimize negative impact on existing site characteristics and operations.
- Ensure the targeted quality level is met, resulting in satisfied customers and less future maintenance.
- Ensure proper measures are taken to prevent accidents.

Partnering can foster a mindset among project participants to proactively identify and accommodate shared values in order to achieve amicable working relationships. This allows partners the tools they need to resolve disputes through compromise. Partnering empowers lower levels of management to accept responsibility and problem-solving authority. Adopting partnering principles—and integrating them into project authorization and execution processes—builds beneficial relationships.

## Characteristics Required for Successful Partnering

Each member of a successful partnering process must possess the following:

- Ethical conduct
- Incentive to achieve mutually established goals
- Positive attitude
- Good interpersonal skills
- Willingness to compromise, when necessary, and maintain fairness
- Knowledge, skills, and ability to do the job

To demonstrate the positive attitude and good interpersonal skills involved in partnering, a facility owner may pass along congratulatory emails or letters of appreciation from clients and end-users to contractors and designers, as acknowledgment of their quality of performance.

Willingness to compromise with fairness can be displayed when problems arise during the course of construction. For example, owners can work with contractors to accommodate materials deliveries as close as possible to the site during times that will have the least impact on operations. Owners can also readily convey information about anticipated changes in building-use schedules so that the contractor has time to reallocate labor or subcontracted resources without financial burden.

The contractor in a JOC partnering arrangement might also provide preliminary cost estimates for future projects to assist the owner in establishing budgets—and might offer value engineering suggestions if those estimates exceed available funding.

## How JOC Uniquely Encourages Partnering

When it comes to partnering, JOC is different from all other project delivery methods—even other methods that utilize strategic partnering principles. What sets JOC apart is that it establishes *mutual incentives* for all program participants to openly cooperate and collaborate. As a result, processes are streamlined and obstacles are overcome, making it easier to complete projects on time and within established budgets.

With JOC, the owner has a vested interest in avoiding the time-consuming and costly task of applying DBB methodologies to each small project—and the likelihood of unstable pricing structures. The contractor has a vested interest in providing the owner with quality service in order to continue to receive additional work. In fact, it is the facility owner's customer who receives the most benefit from JOC partnering.

Unlike most other construction delivery methods, JOC encourages the establishment of beneficial relationships that are long-term, rather than for the duration of one project only. JOC contracts can cover multiple projects of varying scale and scope over extended periods of time. Mutual respect, trust, consistency, and a joint commitment to achieve defined goals—these are all attributes of long-term beneficial relationships. JOC reaches its optimum benefit for all participants when these attributes are consistently present over time.

In a JOC arrangement, some projects will be completed as new ones are started. This promotes collaboration, allows an opportunity to enhance processes, and builds mutually beneficial long-term relationships among program participants (as shown below).

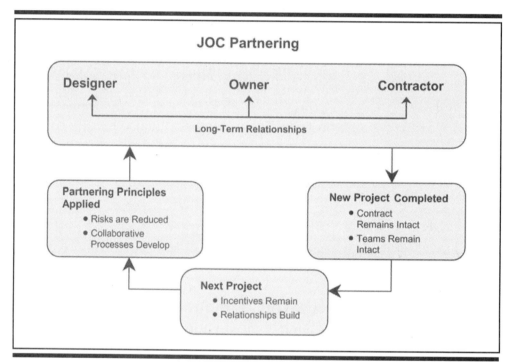

**Figure 8.1** shows strategic partnering principles applied to JOC, which are complementary to the method's inherent attributes. Key participants' mutual incentives are aligned to program success based on vested interests. Also, note the cyclic nature of the method, capable of handling high volumes of work simultaneously while promoting efficiency as each new project is completed.

## Conclusion

Open communication and collaborative processes begin to develop during the program start-up meeting. Performance benchmarking and any necessary adjustments can be made as a result of issues raised at progress meetings. This teamwork atmosphere continues during the JO scoping process as the owner provides detailed information about existing site and job conditions (if available) and works closely with the contractor to schedule the project for the least impact on the facility's operations. The contractor offers technical assistance and openly suggests value engineering methods.

Since pricing structures are pre-established at the beginning of the contract term, the contractor can focus on performance rather than the bottom line. The end result is synergetic in nature—each component collaborative process is greater as a part of the whole process.

A successful JOC program exhibits compromise with fairness throughout the development of the JO proposal. As all of its components are mutually agreed to, the JO proposal is approved, reducing the likelihood for dispute later on. Once the JO is authorized for accomplishment, all key participants have a vested interest to ensure that each project is satisfactorily completed on time and on budget.

JOC and strategic partnering go hand-in-hand. A look behind the scenes of proper contract execution will reveal an atmosphere of cooperative collaboration among key participants. Although consistent commitment to mutually defined goals is not always easy to maintain, with JOC it is easier to accomplish and well worth the effort.

1. Barlow et al. 1997.

2. Construction Industry Institute Australia, 1996.

# Appendix

The sample JOC RFP Evaluation Summary form can help proposal reviewers assess and select successful contract awardee(s). The form allows for comparison of up to five reviewers' evaluation scores. A separate evaluation form would need to be generated for each RFP respondent. Evaluation categories and associated weightings are at the owner's discretion and may vary as needed to meet owner requirements. Note that a HUB plan evaluation is not included, as some owners may not have a corresponding requirement, and others may identify the submittal of a HUB plan not as a weighted score, but as a requirement for consideration of the proposal.

This type of checklist can be useful for both the owner and the contractor – as a reminder of key issues to address with attendees during site visits. The items listed here commonly arise during the JO scoping process. (The checklist shown is an example only. Actual project requirements may vary.)

RSMeans is the most widely used and accepted source of cost data among JOC practitioners. Several pages from Means' *Facilities Construction Cost Data 2005* are included in this Appendix to assist in understanding cost estimating technologies and basic processes. These pages include an overview of Means cost data, including how costs for labor, materials, equipment, general conditions, and overhead and profit are calculated. Basic guidelines on how to read the unit-price data follow. For more information on Means cost data, visit www.rsmeans.com.

Name of Respondent: _____

RFP Number: _____

Date of Evaluation: _____

Rate each item from 1 (poor) to 10 (excellent).

| JOC RFP Evaluation Summary | Reviewers' | Rating 1 | Rating 2 | Rating 3 | Rating 4 | Rating 5 | Total Rating | Average Rating | Item Weight | Total Rating | Category Weight | Final Score |
|---|---|---|---|---|---|---|---|---|---|---|---|---|
| **a.) Respondent's Compared CPP** | | | | | | | | | | | 20% | 0.00 |
| | | (Lowest CPP/this CPP) × (100 × 20) = Final Score This Category) | | | | | | | | | | |
| **b.) Past Performance & Relevant Experience** | | | | | | | | | | | | |
| 1. Performance reference check on previous contracts | | 0+ | 0+ | 0+ | 0+ | 0= | 0 | /5= | 0 × | 40%= | 0.00 | |
| 2. Meets years-in-business requirements | | 0+ | 0+ | 0+ | 0+ | 0= | 0 | /5= | 0 × | 25%= | 0.00 | |
| 3. Projects of similar value, scope, and scale | | 0+ | 0+ | 0+ | 0+ | 0= | 0 | /5= | 0 × | 5%= | 0.00 | |
| 4. Performed work simultaneously on multiple projects | | 0+ | 0+ | 0+ | 0+ | 0= | 0 | /5= | 0 × | 20%= | 0.00 | |
| 5. Partnering and teamwork experience | | 0+ | 0+ | 0+ | 0+ | 0= | 0 | /5= | 0 × | 10%= | 0.00 | |
| | | | | | | | | | | 100% | 20% | 0.00 |
| **c.) Project Management/Technical Ability** | | | | | | | | | | | | |
| 6. Key project management staff | | 0+ | 0+ | 0+ | 0+ | 0= | 0 | /5= | 0 × | 15%= | 0.00 | |
| 7. Technical support staff | | 0+ | 0+ | 0+ | 0+ | 0= | 0 | /5= | 0 × | 10%= | 0.00 | |
| 8. Quality control plan | | 0+ | 0+ | 0+ | 0+ | 0= | 0 | /5= | 0 × | 20%= | 0.00 | |
| 9. Financial capability | | 0+ | 0+ | 0+ | 0+ | 0= | 0 | /5= | 0 × | 15%= | 0.00 | |
| 10. Execution and schedule plans | | 0+ | 0+ | 0+ | 0+ | 0= | 0 | /5= | 0 × | 15%= | 0.00 | |
| 11. Overall ability to coordinate multiple projects | | 0+ | 0+ | 0+ | 0+ | 0= | 0 | /5= | 0 × | 10%= | 0.00 | |
| 12. Safety and environmental control plans | | 0+ | 0+ | 0+ | 0+ | 0= | 0 | /5= | 0 × | 5%= | 0.00 | |
| 13. Partnering and team building plan | | 0+ | 0+ | 0+ | 0+ | 0= | 0 | /5= | 0 × | 10%= | 0.00 | |
| | | | | | | | | | | 100% | 25% | 0.00 |
| **d.) Subcontractor Management and Materials Support Capability** | | | | | | | | | | | | |
| 14. Subcontractor management plan | | 0+ | 0+ | 0+ | 0+ | 0= | 0 | /5= | 0 × | 25%= | 0.00 | |
| 15. Identification of key subcontractors and material suppliers | | 0+ | 0+ | 0+ | 0+ | 0= | 0 | /5= | 0 × | 25%= | 0.00 | |
| 16. Purchasing system/level of subcontracting | | 0+ | 0+ | 0+ | 0+ | 0= | 0 | /5= | 0 × | 30%= | 0.00 | |
| 17. Subcontractor assistance plan | | 0+ | 0+ | 0+ | 0+ | 0= | 0 | /5= | 0 × | 10%= | 0.00 | |
| 18. Plan for obtaining professional consulting services | | 0+ | 0+ | 0+ | 0+ | 0= | 0 | /5= | 0 × | 10%= | 0.00 | |
| | | | | | | | | | | 100% | 15% | 0.00 |
| **e.) Sample Job Order Proposal** | | | | | | | | | | | | |
| 19. Code review drawing | | 0+ | 0+ | 0+ | 0+ | 0= | 0 | /5= | 0 × | 20%= | 0.00 | |
| 20. Scope of work | | 0+ | 0+ | 0+ | 0+ | 0= | 0 | /5= | 0 × | 40%= | 0.00 | |
| 21. JO price proposal | | 0+ | 0+ | 0+ | 0+ | 0= | 0 | /5= | 0 × | 40%= | 0.00 | |
| | | | | | | | | | | 100% | 15% | 0.00 |
| **f.) Interview Results** | | 0+ | 0+ | 0+ | 0+ | 0= | 0 | /5= | 0 × | 100%= | 0.00 | |
| | | | | | | | | | | 100% | 5% | 0.00 |
| | | | | | | | | | | | Total Score | 0.00 |

# PRE-CONSTRUCTION CHECKLIST

Project Name: _____    Date of Site Visit: _____

Project or Site Location: _____

Job Order #: _____

Project Start Date: _____    Project Completion Date: _____

Site Visit Attendees/Names and Contact Information:

Owner's Representative: _____

Contractor's Representative: _____

Professional Consultant: _____

Other Attendees: _____

## *Agenda Items, as Applicable:*

☐ **Required Insurance, Bonding, HUB Plan**

☐ **Performance Dates and Times**

☐ **Building Usage and Operations Issues, Limits of Construction, Record Documents, Design Documents**

☐ **Required Job Site Postings**

☐ **Staging, Sequence Schedules, Lead Times for Specialty Items**

☐ **Access/Access Times, Temporary Facilities, Equipment Usage**

☐ **Key/Card Issuance and Return Procedures**

☐ **Signage, Barricades**

☐ **Parking, Deliveries**

☐ **Presence of Hazardous Materials, Other Regulatory Requirements**

☐ **Indoor/Outdoor Environmental Protection (Fumes, Dust, Etc.)**

☐ **Worker Behavior, Job Site Safety, Emergency Procedures**

☐ **Inspections and Quality Control**

☐ **Building or Site Protection, Construction Aids**

☐ **Field Orders, Change Orders, Requests for Information**

☐ **Submittals:**          ☐ **Permits:**

    ☐ **Shop Drawings**          ☐ **Building**

    ☐ **Field Samples**          ☐ **Confined Space Entry**

    ☐ **Mock-ups**          ☐ **Hot Work**

    ☐ **Specified Materials**          ☐ **Trenching**

    ☐ **MSDS**          ☐ **"One Call" to Identify Underground Utilities**

    ☐ **Other** _____          ☐ **Other** _____

☐ **Other Agenda Items:** _____

# *How the Book Is Built: An Overview*

## *A Powerful Construction Tool*

You have in your hands one of the most powerful construction tools available today. A successful project is built on the foundation of an accurate and dependable estimate. This book will enable you to construct just such an estimate.

For the casual user the book is designed to be:

- quickly and easily understood so you can get right to your estimate.
- filled with valuable information so you can understand the necessary factors that go into the cost estimate.

For the regular user, the book is designed to be:

- a handy desk reference that can be quickly referred to for key costs.
- a comprehensive, fully reliable source of current construction costs and productivity rates, so you'll be prepared to estimate any project.
- a source book for preliminary project cost, product selections, and alternate materials and methods.

To meet all of these requirements we have organized the book into the following clearly defined sections.

## How To Use the Book: The Details

This section contains an in-depth explanation of how the book is arranged . . . and how you can use it to determine a reliable construction cost estimate. It includes information about how we develop our cost figures and how to completely prepare your estimate.

## Unit Price Section

All cost data has been divided into the 16 divisions according to the MasterFormat system of classification and numbering as developed by the Construction Specifications Institute (CSI) and Construction Specifications Canada (CSC). For a listing of these divisions and an outline of their subdivisions, see the Unit Price Section Table of Contents.

*Estimating tips are included at the beginning of each division.*

### Division 17: Quick Project Estimates

In addition to the 16 Unit Price Divisions there is a S.F. (Square Foot) and C.F. (Cubic Foot) Cost Division, Division 17. It contains costs for 59 different building types that allow you to quickly make a rough estimate for the overall cost of a project or its major components.

## Assemblies Section

The cost data in this section has been organized in an "Assemblies" format. These assemblies are the functional elements of a building and are arranged according to the 7 divisions of the UNIFORMAT II classification system. For a complete explanation of a typical "Assemblies" page, see "How To Use the Assemblies Cost Tables."

## Reference Section

This section includes information on Reference Numbers, Change Orders, Crew Listings, Historical Cost Indexes, City Cost Indexes, and Location Factors and a listing of Abbreviations. It is visually identified by a vertical gray bar on the page edges.

**Reference Numbers:** At the beginning of selected major classifications throughout the book are "reference numbers" shown in bold squares. These numbers refer you to related information in other sections.

In these other sections, you'll find related tables, explanations, and estimating information. Also included are alternate pricing methods and technical data, along with information on design and economy in construction. You'll also find helpful tips on estimating and construction.

**It is recommended that you refer to this Reference Section if a "reference number" appears within the section you are estimating.**

**Change Orders:** This section includes information on the factors that influence the pricing of change orders.

**Crew Listings:** This section lists all the crews referenced. For the purposes of this book, a crew is composed of more than one trade classification and/or the addition of equipment to any trade classification. Power equipment is included in the cost of the crew. Costs are shown both with bare labor rates and with

the installing contractor's overhead and profit added. For each, the total crew cost per eight-hour day and the composite cost per labor-hour are listed.

**Historical Cost Indexes:** These indexes provide you with data to adjust construction costs over time. If you know costs for a past project, you can use these indexes to estimate the cost to construct the same project today.

**City Cost Indexes:** Costs vary depending on the regional economy. You can adjust the national average costs in this book to over 930 locations throughout the U.S. and Canada by using the data in this section. How to use information is included.

*Location Factors, to quickly adjust the data to over 930 zip code areas, are included.*

**Abbreviations:** A listing of the abbreviations and the terms they represent is included.

## Index

A comprehensive listing of all terms and subjects in this book will help you quickly find what you need

## The Scope of This Book

This book is designed to be comprehensive and easy to use. To that end we have made certain assumptions:

1. We have established material prices based on a national average.
2. We have computed labor costs based on a 30-city national average of union wage rates.
3. We have targeted the data for projects of a certain size range.

For a more detailed explanation of how the cost data is developed, see "How To Use the Book: The Details."

## Project Size

This book is aimed primarily at commercial and industrial projects costing $10,000 to $1,000,000.

**With reasonable exercise of judgment the figures can be used for any building work.** *However, for civil engineering structures such as bridges, dams, highways, or the like, please refer to Means Heavy Construction Cost Data.*

# How to Use the Book: The Details

## What's Behind the Numbers? The Development of Cost Data

The staff at RSMeans continuously monitors developments in the construction industry in order to ensure reliable, thorough and up-to-date cost information.

While *overall* construction costs may vary relative to general economic conditions, price fluctuations within the industry are dependent upon many factors. Individual price variations may, in fact, be opposite to overall economic trends. Therefore, costs are continually monitored and complete updates are published yearly. Also, new items are frequently added in response to changes in materials and methods.

## Costs—$ (U.S.)

All costs represent U.S. national averages and are given in U.S. dollars. The Means City Cost Indexes can be used to adjust costs to a particular location. The City Cost Indexes for Canada can be used to adjust U.S. national averages to local costs in Canadian dollars. No exchange rate conversion is necessary.

## Material Costs

The RSMeans staff contacts manufacturers, dealers, distributors, and contractors all across the U.S. and Canada to determine national average material costs. If you have access to current material costs for your specific location, you may wish to make adjustments to reflect differences from the national average. Included within material costs are fasteners for a normal installation. RSMeans engineers use manufacturers' recommendations, written specifications, and/or standard construction practice for size and spacing of fasteners. Adjustments to material costs may be required for your specific application or location. Material costs do not include sales tax.

## Labor Costs

Labor costs are based on the average of wage rates from 30 major U.S. cities. Rates are determined from labor union agreements or prevailing wages for construction trades for the current year. Rates, along with overhead and profit markups, are listed on the inside back cover of this book.

- If wage rates in your area vary from those used in this book, or if rate increases are expected within a given year, labor costs should be adjusted accordingly.

Labor costs reflect productivity based on actual working conditions. These figures include time spent during a normal workday on tasks other than actual installation, such as material receiving and handling, mobilization at site, site movement, breaks, and cleanup.

Productivity data is developed over an extended period so as not to be influenced by abnormal variations and reflects a typical average.

## Equipment Costs

Equipment costs include not only rental, but also operating costs for equipment under normal use. The operating costs include parts and labor for routine servicing such as repair and replacement of pumps, filters, and worn lines. Normal operating expendables, such as fuel, lubricants, tires, and electricity (where applicable), are also included. Extraordinary operating expendables with highly variable wear patterns, such as diamond bits and blades, are excluded. These costs are included under materials. Equipment rental rates are obtained from industry sources throughout North America—contractors, suppliers, dealers, manufacturers, and distributors.

**Crew Equipment Cost/Day**—The power equipment required for each crew is included in the crew cost. The daily cost for crew equipment is based on dividing the weekly bare rental rate by 5 (number of working days per week), and then adding the hourly operating cost times 8 (hours per day). This "Crew Equipment Cost/Day" is listed in Subdivision 01590.

**Mobilization/Demobilization**—The cost to move construction equipment from an equipment yard or rental company to the job site and back again is not included in equipment costs. Mobilization (to the site) and demobilization (from the site) costs can be found in Section 02305-250. If a piece of equipment is already at the job site, it is not appropriate to utilize mob/demob costs again in an estimate.

## General Conditions

Cost data in this book is presented in two ways: Bare Costs and Total Cost including O&P (Overhead and Profit). General Conditions, when applicable, should also be added to the Total Cost including O&P. The costs for General Conditions are listed in Division 1 of the Unit Price Section and the Reference Section of this book. General Conditions for the *Installing Contractor* may range from 0% to 10% of the Total Cost including O&P. For the *General* or *Prime Contractor*, costs for General Conditions may range from 5% to 15% of the Total Cost including O&P, with a figure of 10% as the most typical allowance.

## Overhead and Profit

Total Cost, including O&P, for the *Installing Contractor* is shown in the last column on both the Unit Price and Assemblies pages of this book. This figure is the sum of the bare material cost plus 10% for profit, the base labor cost plus total overhead and profit, and the bare equipment cost plus 10% for profit. Details for the calculation of Overhead and Profit on labor are shown on the inside back cover and in the Reference Section of this book. (See the "How to Use the Unit Price Pages" for an example of this calculation.)

## Factors Affecting Costs

Costs can vary depending upon a number of variables. Here's how we have handled the main factors affecting costs.

**Quality**—The prices for materials and the workmanship upon which productivity is based represent sound construction work. They are also in line with U.S. government specifications.

**Overtime**—We have made no allowance for overtime. If you anticipate premium time or work beyond normal working hours, be sure to make an appropriate adjustment to your labor costs.

**Productivity**—The productivity, daily output, and labor-hour figures for each line item are based on working an eight-hour day in daylight hours in moderate temperatures. For work that extends beyond normal work hours or is performed under adverse conditions, productivity may decrease. (See the section in "How To Use the Unit Price Pages" for more on productivity.)

**Size of Project**—The size, scope of work, and type of construction project will have a significant impact on cost. Economies of scale can reduce costs for large projects. Unit costs can often run higher for small projects. Costs in this book are intended for the size and type of project as previously described in "How the Book Is Built: An Overview." Costs for projects of a significantly different size or type should be adjusted accordingly.

**Location**—Material prices in this book are for metropolitan areas. However, in dense urban areas, traffic and site storage limitations may increase costs. Beyond a 20-mile radius of large cities, extra trucking or transportation charges may also increase the material costs slightly. On the other hand, lower wage rates may be in effect. Be sure to consider both of these factors when preparing an estimate, particularly if the job site is located in a central city or remote rural location.

In addition, highly specialized subcontract items may require travel and per-diem expenses for mechanics.

**Other Factors**—
- season of year
- contractor management
- weather conditions
- local union restrictions
- building code requirements
- availability of:
  - adequate energy
  - skilled labor
  - building materials
- owner's special requirements/restrictions
- safety requirements
- environmental considerations

**Unpredictable Factors**—General business conditions influence "in-place" costs of all items. Substitute materials and construction methods may have to be employed. These may affect the installed cost and/or life cycle costs. Such factors may be difficult to evaluate and cannot necessarily be predicted on the basis of the job's location in a particular section of the country. Thus, where these factors apply, you may find significant, but unavoidable cost variations for which you will have to apply a measure of judgment to your estimate.

## Rounding of Costs

In general, all unit prices in excess of $5.00 have been rounded to make them easier to use and still maintain adequate precision of the results. The rounding rules we have chosen are in the following table.

| Prices from ... | Rounded to the nearest ... |
|---|---|
| $.01 to $5.00 | $.01 |
| $5.01 to $20.00 | $.05 |
| $20.01 to $100.00 | $.50 |
| $100.01 to $300.00 | $1.00 |
| $300.01 to $1,000.00 | $5.00 |
| $1,000.01 to $10,000.00 | $25.00 |
| $10,000.01 to $50,000.00 | $100.00 |
| $50,000.01 and above | $500.00 |

## Important Estimating Considerations

One reason for listing a job size "minimum" is to ensure that the construction craftsmen are productive for 8 hours a day on a continuous basis. Otherwise, tasks that require only 6 to 7 hours could be billed as an 8-hour day, thus providing an erroneous productivity estimate. The "productivity," or daily output, of each craftsman includes mobilization and cleanup time, break time, and plan layout time, as well as an allowance to carry stock from the storage trailer or location on the job site up to 200'

into the building and onto the first or second floor. If material has to be transported over greater distances or to higher floors, an additional allowance should be considered by the estimator. An allowance has also been included in the piping and fittings installation time for a leak check and minor tightening.

Equipment installation time includes the following applicable items: positioning, leveling and securing the unit in place, connecting all associated piping, ducts, vents, etc., which shall have been estimated separately, connecting to an adjacent power source, filling/bleeding, startup, adjusting the controls up and down to ensure proper response, setting the integral controls/valves/regulators/thermostats for proper operation (does not include external building type control systems, DDC systems, etc.), explaining/training owner's operator, and warranty. A reasonable breakdown of the labor costs is as follows:

1. Movement into building, installation/ setting of equipment — 35%
2. Connection to piping/duct/power, etc. — 25%
3. Filling/flushing/cleaning/touchup, etc. — 15%
4. Startup/running adjustments — 5%
5. Training owner's rep. — 5%
6. Warranty/call back/service — 15%

Note that cranes or other lifting equipment are not included on any lines in Division 15. For example, if a crane is required to lift a heavy piece of pipe into place high above a gym floor, or to put a rooftop unit on the roof of a four-story building, etc., it must be added. Due to the potential for extreme variation—from nothing additional required to a major crane or helicopter—we feel that including a nominal amount for "lifting contingency" would be useless and detract from the accuracy of the estimate. When using equipment rental from Means, remember to include the cost of the operator(s).

# How to Use the Unit Price Pages

*The following is a detailed explanation of a sample entry in the Unit Price Section. Next to each bold number below is the described item with the appropriate component of the sample entry following in parentheses. Some prices are listed as bare costs, others as costs that include overhead and profit of the installing contractor. In most cases, if the work is to be subcontracted, the general contractor will need to add an additional markup (RSMeans suggests using 10%) to the figures in the column "Total Incl. O&P."*

## 1 Division Number/Title (03300/Cast-In-Place Concrete)

Use the Unit Price Section Table of Contents to locate specific items. The sections are classified according to the CSI MasterFormat (1995 Edition).

## 2 Line Numbers (03310 240 3900)

Each unit price line item has been assigned a unique 12-digit code based on the CSI MasterFormat classification.

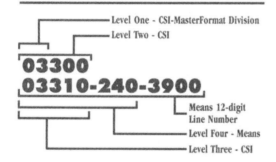

- Level One - CSI-MasterFormat Division
- Level Two - CSI

**03300**
**03310-240-3900**

- Means 12-digit Line Number
- Level Four - Means
- Level Three - CSI

## 3 Description (Concrete-In-Place, etc.)

Each line item is described in detail. Sub-items and additional sizes are indented beneath the appropriate line items. The first line or two after the main item (in boldface) may contain descriptive information that pertains to all line items beneath this boldface listing.

## 4 Reference Number Information

| R03310 -010 |

You'll see reference numbers shown in bold rectangles at the beginning of some sections. These refer to related items in the Reference Section, visually identified by a vertical gray bar on the page edges.

The relation may be: (1) an estimating procedure that should be read before estimating, (2) an alternate pricing method, or (3) technical information.

The "R" designates the Reference Section. The numbers refer to the MasterFormat classification system.

**It is strongly recommended that you review all reference numbers that appear within the section in which you are working.**

Note: Not all reference numbers appear in all Means publications.

---

## 03300 | Cast-In-Place Concrete

### 03310 | Structural Concrete

| | | | CREW | DAILY OUTPUT | LABOR-HOURS | UNIT | MAT. | LABOR | EQUIP. | TOTAL | TOTAL INCL O&P | |
|---|---|---|---|---|---|---|---|---|---|---|---|---|
| **240** | 0010 | **CONCRETE IN PLACE** Including forms (4 uses), reinforcing | R03310 -010 | | | | | | | | | **240** |
| | 0050 | steel and finishing unless otherwise indicated | | | | | | | | | | |
| | 0300 | Beams, 5 kip per L.F., 10' span | R03310 -100 | C-14A | 15.62 | 12.804 | C.Y. | 284 | 440 | 46.50 | 770.50 | 1,100 | |
| | 0350 | 25' span | | | 18.55 | 10.782 | | 214 | 370 | 39.50 | 623.50 | 895 | |
| | 3500 | Add per floor for 3 to 6 stories high | | | 31,800 | .002 | | | .05 | .02 | .07 | .11 | |
| | 3520 | For 7 to 20 stories high | | | 21,200 | .003 | | | .08 | .03 | .11 | .17 | |
| | 3800 | Footings, spread under 1 C.Y. | | C-14C | 38.07 | 2.942 | | 162 | 96 | .62 | 258.62 | 340 | |
| | 3850 | Over 5 C.Y. | | | 81.04 | 1.382 | | 226 | 45 | .29 | 271.29 | 325 | |
| | 3900 | Footings, strip, 18" x 9", unreinforced | | | 40 | 2.800 | | 101 | 91.50 | .59 | 193.09 | 264 | |
| | 3920 | 18" x 9", reinforced | | | 35 | 3.200 | | 120 | 105 | .67 | 225.67 | 305 | |
| | 3925 | 20" x 10", unreinforced | | | 45 | 2.489 | | 98 | 81.50 | .52 | 180.02 | 244 | |
| | 3930 | 20" x 10", reinforced | | | 40 | 2.800 | | 114 | 91.50 | .59 | 206.09 | 278 | |
| | 3935 | 24" x 12", unreinforced | | | 55 | 2.036 | | 97 | 66.50 | .43 | 163.93 | 217 | |
| | 3940 | 24" x 12", reinforced | | | 48 | 2.333 | | 113 | 76 | .49 | 189.49 | 252 | |

## Crew (C-14C)

The "Crew" column designates the typical trade or crew used to install the item. If an installation can be accomplished by one trade and requires no power equipment, that trade and the number of workers are listed (for example, "2 Carpenters"). If an installation requires a composite crew, a crew code designation is listed (for example, "C-14C"). You'll find full details on all composite crews in the Crew Listings.

* For a complete list of all trades utilized in this book and their abbreviations, see the inside back cover.

### Crews

| Crew No. | Bare Costs | | Incl. Subs O & P | | Cost Per Labor-Hour | |
|---|---|---|---|---|---|---|
| Crew C-14C | Hr. | Daily | Hr. | Daily | Bare Costs | Incl. O&P |
| 1 Carpenter Foreman (out) | $36.25 | $290.00 | $60.05 | $480.40 | $32.66 | $54.26 |
| 6 Carpenters | 34.25 | 1644.00 | 56.75 | 2724.00 | | |
| 2 Rodmen (rein't.) | 37.95 | 607.20 | 65.25 | 1044.00 | | |
| 4 Laborers | 26.70 | 854.40 | 44.25 | 1416.00 | | |
| 1 Cement Finisher | 32.85 | 262.80 | 51.65 | 413.20 | | |
| 1 Gas Engine Vibrator | | 24.00 | | 26.40 | .21 | .24 |
| 112 L.H., Daily Totals | | $3682.40 | | $6104.00 | $32.87 | $54.50 |

## Productivity: Daily Output (40)/Labor-Hours (2.80)

The "Daily Output" represents the typical number of units the designated crew will install in a normal 8-hour day. To find out the number of days the given crew would require to complete the installation, divide your quantity by the daily output. For example:

| Quantity | ÷ | Daily Output | = | Duration |
|---|---|---|---|---|
| 100 C.Y. | ÷ | 40/ Crew Day | = | 2.50 Crew Days |

The "Labor-Hours" figure represents the number of labor-hours required to install one unit of work. To find out the number of labor-hours required for your particular task, multiply the quantity of the item times the number of labor-hours shown. For example:

| Quantity | x | Productivity Rate | = | Duration |
|---|---|---|---|---|
| 100 C.Y. | x | 2.800 Labor-Hours/ C.Y. | = | 280 Labor-Hours |

## Unit (C.Y.)

The abbreviated designation indicates the unit of measure upon which the price, production, and crew are based (C.Y. = Cubic Yard). For a complete listing of abbreviations refer to the Abbreviations Listing in the Reference Section of this book.

## Bare Costs:

### Mat. (Bare Material Cost) (101)

The unit material cost is the "bare" material cost with no overhead or profit included. *Costs shown reflect national average material prices for January of the current year and include delivery to the job site. No sales taxes are included.*

### Labor (91.50)

The unit labor cost is derived by multiplying bare labor-hour costs for Crew C-14C by labor-hour units. The bare labor-hour cost is found in the Crew Section under C-14C. (If a trade is listed, the hourly labor cost—the wage rate—is found on the inside back cover.)

| Labor-Hour Cost Crew C-14C | x | Labor-Hour Units | = | Labor |
|---|---|---|---|---|
| $32.66 | x | 2.80 | = | $91.45 |

### Equip. (Equipment) (.59)

Equipment costs for each crew are listed in the description of each crew. Tools or equipment whose value justifies purchase or ownership by a contractor are considered overhead as shown on the inside back cover. The unit equipment cost is derived by multiplying the bare equipment hourly cost by the labor-hour units.

| Equipment Cost Crew C-14C | x | Labor-Hour Units | = | Equip. |
|---|---|---|---|---|
| .21 | x | 2.80 | = | .59 |

### Total (193.09)

The total of the bare costs is the arithmetic total of the three previous columns: mat., labor, and equip.

| Material | + | Labor | + | Equip. | = | Total |
|---|---|---|---|---|---|---|
| $101 | + | $91.50 | + | $.59 | = | $193.09 |

## Total Costs Including O&P

This figure is the sum of the bare material cost plus 10% for profit; the bare labor cost plus total overhead and profit (per the inside back cover or, if a crew is listed, from the crew listings); and the bare equipment cost plus 10% for profit.

| | | |
|---|---|---|
| Material is Bare Material Cost + 10% = 101 + 10.10 | = | $111.10 |
| Labor for Crew C-14C = Labor-Hour Cost (54.26) x Labor-Hour Units (2.80) | = | $151.93 |
| Equip. is Bare Equip. Cost + 10% = .59 + .06 | = | $ .65 |
| Total (Rounded) | = | $264 |

# Resources

## Organizations

**American Institute of Architects (AIA)**
1735 New York Avenue, NW
Washington, DC 20006-5292
www.aia.org

**Associated General Contractors of America (AGC)**
333 John Carlyle Street
Suite 200
Alexandria, VA 22314
www.agc.org

**Association for Facilities Engineering (AFE)**
8160 Corporate Park Drive
Suite 125
Cincinnati, OH 45242
www.afe.org

**Association of Higher Education Facilities Officers (APPA)**
1643 Prince Street
Alexandria, VA 22314-2818
www.appa.org

**Center for Job Order Contracting Excellence (CJE)**
Alliance for Construction Excellence (ACE)
Del E. Webb School of Construction
Arizona State University
P.O. Box 870204
Tempe, AZ 85287-0204
http://construction.asu.edu/ace/cje.htm

**Construction Industry Institute (CII)**
3925 West Breaker Lane
Austin, TX 78759-5316
www.construction-institute.org

**Construction Specifications Canada (CSC)**
120 Carlton Street
Suite 312
Toronto, ON M5A 4K2
www.csc-dcc.ca

**Construction Specifications Institute (CSI)**
99 Canal Center Plaza
Suite 300
Alexandria, VA 22314
www.csinet.org

**International Facility Management Association (IFMA)**
1 E. Greenway Plaza
Suite 1100
Houston, TX 77046
www.ifma.org

**RSMeans Business Solutions**
63 Smiths Lane
Kingston, MA 02364-0800
www.rsmeans.com

**Texas Job Order Contract Conference (TJOCC)**
P.O. Box 1241
Wimberley, TX 78676-1241
www.tjocc.org

**Texas State University-San Marcos**
Facilities Planning, Design, and Construction
601 University Drive
San Marcos, TX 78666
www.txstate.edu

**The Cooperative Purchasing Network (TCPN)**
7145 W. Tidwell Road
Houston, TX 77092
www.tcpn.org

**The U.S. Army Corps of Engineers (USACE)**
441 G. Street, NW
Washington, DC 20314
www.usace.army.mil

**The U.S. Department of Energy (DOE)**
1000 Independence Avenue, SW
Washington, DC 20585
www.energy.gov

**The Wool-Zee Co., Inc.**
4200 Meridian Street
Suite 216
Bellingham, WA 98226
360-676-4604
www.woolzee.com

## Publications

*Architectural Graphic Standards,* John Wiley & Sons, Inc.

Associated General Contractors of America (AGC), *Partnering: A Concept for Success,* AGC 1991.

Barlow, J., Cohen, M., Jashapara, A., and Simpson, Y., *Towards Positive Partnering,* Bristol: The Policy Press, 1997.

Bramble, Barry B. and West, Joseph D., *Design-Build Contracting Claims, 2003 Cumulative Supplement,* Construction Law Library, Aspen Publishers, Inc., New York, 2003.

Center of Job Order Contracting Excellence, *CJE Newsletter,* http://construction.asu.edu/ace/cje.htm.

Collier, Keith, *Construction Contracts, 3rd edition,* Prentice-Hall, Inc., Upper Saddle River, New Jersey, 2001.

Construction Industry Institute Australia (CIIA), *Partnering: Models for Success,* Research Report No. 8, Sydney: CIIA. 1996.

Fisk, Edward R., *Construction Project Administration, 7th edition,* Pearson Education, Inc., Upper Saddle River, New Jersey, 2003.

Hauf, Harold D., *Building Contracts for Design & Construction, 2nd edition,* John Wiley & Sons, Inc., 1968.

Henderson, A., *Delivery Order Construction for Small Colleges. The Proceedings: 1996 Educational Conference & 83rd Annual Meeting,* APPA: The Association of Higher Education Facilities Officers, 1996.

United States Army, *Job Order Contracting Guide,* January, 2003. http://www.hqda.army.mil/acsimweb/fd/docs/JOC%20Guide.pdf

*Note: The publisher and the author do not necessarily promote or recommend the above resources. They are provided for your information only.*

**RSMeans offers a number of useful reference and estimating books** *(see www.rsmeans.com for a complete list):*

*Facilities Construction Cost Data,* RSMeans, 2005.*

*Facilities Maintenance & Repair Cost Data,* RSMeans, 2005.*

*Repair & Remodeling Cost Data,* RSMeans, 2005.*

*ADA Compliance Pricing Guide, 2nd edition,* by Adaptive Environments Center, RSMeans, 2004.

*Building Professional's Guide to Contract Documents, 3rd edition,* by Waller Poage, RSMeans, 2000.

*Cost Planning and Estimating for Facilities Maintenance,* RSMeans, 1996.

*Electrical Estimating Methods, 3rd edition,* RSMeans, 2003.

*Environmental Remediation Estimating Methods, 2nd edition,* by Richard R. Rast, RSMeans, 2003.

*Facilities Operations & Engineering Reference,* RSMeans, 1999.

*Historic Preservation: Project Planning & Estimating,* by Swanke Hayden Connell Architects, RSMeans, 2000.

*Landscape Estimating Methods, 4th edition,* by Sylvia Fee, RSMeans, 2002.

*Mechanical Estimating Methods, 3rd edition,* by Edward Wetherill and Merl Vandervort, RSMeans, 2002.

*Plumbing Estimating Methods, 3rd edition,* by Joseph Galeno and Sheldon Greene, RSMeans, 2004.

*Repair & Remodeling Estimating Methods, 4th edition,* RSMeans, 2002.

**\*Cost data books are also available electronically as *CostWorks.***

**RSMeans JOC services include:**

- Development of JOC documents including the unit-price book, technical specifications, contract terms and conditions, and bid documents, including:
  - Catalog of construction tasks
  - Standard set of technical specifications
  - JO contractual terms and conditions
  - Model set of JOC execution procedures and policies
  - Model JOC management software including contractor cost proposals, cost estimates, and management reports and forms
- Activities needed to establish the structure of the JOC program, such as informing the internal staff and the contracting community about JOC

- Assistance with procurement of JOC contractors, development and coordination of JOC execution procedures and documents, and training, including:
  - JOC program structure/bidding strategy
  - Pre-bid seminars
  - Development of JOC proposals
  - External and internal marketing
- *Means JOCWorks*™—A new JOC-specific estimating and project management program. Features include:
  - Import/export of estimates by disk or email
  - Exportable reports to *Microsoft Excel*™, *Microsoft Word*™, or *Adobe*™ *PDF*
  - Support for multiple and dual award contracts and pricing guides
  - Quick modification tools that allow changes to one or more projects, estimates, or estimate line items
  - Ability to organize projects by contract, contract performance period, contractor, location, estimator, or customer
  - Milestones for tracking projects from design through warranty, including site visits, invoicing, change orders, and material submittals

Visit the RSMeans Web site at **www.rsmeans.com/consulting/joboc.asp** for more information.

# Glossary

**A+B bidding** An arrangement whereby the price proposal is considered in conjunction with a firm's projected time line for the project (typically used with the design-build method of project delivery).

**addendum** A document describing an addition, change, correction, or modification to contract documents. An addendum is issued by the design professional or owner's other designated representative during bidding or prior to the award of contract, and is the primary method of informing bidders of modifications to the work during the bidding process. Addenda become a part of the contract documents.

**advertisement for bids** A published notice of an owner's intention to award a contract for construction to a constructor who submits a proposal according to instructions to bidders. The advertisement is typically published in a convenient form of news media in order to attract constructors who are willing to prepare and submit proposals.

**alternate** A specific item of construction that is set apart by a separate sum. An alternate may or may not be incorporated into the contract sum at the discretion and approval of the owner at the time of contract award. It can increase the base price of the project as an *additive alternate* or decrease it as a *deductive alternate*.

**application for payment** A statement prepared by the contractor stating the amount of work completed and materials purchased and properly stored to date. This statement includes the sum of previous payments and the current payment requested in accordance with payment provisions of the contract documents.

**architect** A design professional who, by education, experience, and examination, is licensed by state government to practice the art of building design and technology.

**as-built drawings** Record drawings made during construction. As-built drawings record the locations, sizes, and nature of concealed items, such as structural elements, accessories, equipment, devices, plumbing lines, valves, mechanical equipment, and the like. These records (with dimensions) form a permanent record for future reference.

**authority having jurisdiction (AHJ)** A person who has the delegated authority to determine, mandate, and enforce code requirements established by jurisdictional governing bodies.

**best overall value** A proposal evaluation process whereby criteria other than pricing is a factor for determining contract award.

**bid** A term commonly used for a complete and properly executed proposal to perform work that has been described in the contract documents and submitted in accordance with instructions to bidders.

**bid bond** A form of bid security purchased by the bidder from a surety. A bid bond is provided, subject to forfeit, to guarantee that the bidder will enter a contract for construction within a specified time and furnish any required bonds, such as performance bond and payment bond.

**bid opening** A formal meeting held at a specified place, date, and time at which sealed bids are opened, tabulated, and read aloud.

**bid security** A bid bond or other form of security, such as a cashier's check, that is acceptable to the owner. Bid security is provided as a guarantee that the bidder will enter into a contract for construction within a specified time and furnish any bonds and other requirements of the bid documents.

**bidding documents** Documents usually including the advertisement or invitation to bidders, instructions to bidders, bid form, form of contract, forms of bonds, conditions of contract, specifications, drawings, and any other information necessary to completely describe the work by which candidate constructors can adequately prepare proposals or bids for the owner's consideration.

**bid shopping** Sharing contractor pricing information among potential bid or proposal respondents prior to award. The practice is considered unethical and sometimes illegal.

**blanket or framework purchase order** A funding commitment that allows authorization for payment for multiple projects from one or more different accounts or from a *clearing* or *revolving account*.

**bonds** Written documents given by a surety in the name of a principal to an obligee to guarantee a specific obligation. In construction, the principal types of bonds are the bid bond, the performance bond, and the payment bond.

**bonus provisions** Provisions in the contract for the construction by which the owner may pay monetary rewards to the contractor for achieving some savings that benefit the owner. For example, a stipulated bond may be offered for early completion of the work or the achievement of some savings in construction cost.

**building codes** The minimum legal requirements established or adopted by a government, such as a municipality. Building codes are established by ordinance and govern the design and construction of buildings.

**building permit** A written authorization required by ordinance for a specific project. A building permit allows construction to proceed in accordance with construction documents approved by the building official.

**cash allowance** A specified sum of money to be included in the contract sum for an element, group of elements, assembly, system, piece of equipment, or other described item to be included in the work under specified conditions.

**certificate of insurance** A written document appropriately signed by a responsible representative of the insurance company and stating the exact coverage and period of time for which the coverage is applicable in accordance with requirements of the contract documents.

**certificate of occupancy** A written document issued by the governing authority in accordance with the provisions of the building permit. It indicates that the project, in the opinion of the building official, has been completed in accordance with applicable building codes. This document gives permission of authorities for the owner to occupy and use the premises for the intended purpose.

**change order** A written document signed by the owner, design professional, and contractor detailing a change or modification to the contract for construction.

**city cost index** Multipliers used to adjust costs to the owner's regional area of operations using percentage ratios.

**clearing or revolving account** A monetary account capable of the continual processing of invoicing and payment transactions.

**code review drawings** Drawings that portray compliance to applicable codes and standards that can be developed by or through the contractor and are supplementary to a written scope of work (also called incidental drawings).

**coefficient** The proposed price multiplier in relation to the pricing structure defined within the JOC proposal documents. The coefficient establishes a competitively bid cost adjustment or multiplier to UPB pricing, taking into consideration the contractor's compensation for services to be rendered to include overhead and profit.

**competitive sealed proposals—stipulated sum** The project-specific agreement between the owner and the contractor to complete the project for a stipulated amount (also called lump-sum or fixed-price contracts.)

**construction drawings** In the contract documents, the graphic representations of the work to be done in the construction project.

**construction manager (CM)** One who directs the process of construction as the agent of either the owner or the contractor, or one who, for a fee, directs and coordinates construction activity carried out under separate or multiple prime contracts.

**construction manager-at-risk (CMAR)** A project-specific delivery method that is suited for medium to large capital or renovation projects. CMAR provides technical assistance to the designer during the design phase, has a cost-capping feature, and allows construction to start before design documents are 100% complete. The CMAR contracts directly with subcontractors, fabricators, and material suppliers.

**construction start-up period** A requirement that may be included in the contract allowing for sequencing of cumulative work volume over a designated time period after contract award.

**consulting engineer** A licensed engineer who may directly contract with the owner for professional services or may be employed by the design professional for the purpose of performing specific tasks of engineering design for portions of a project.

**contingency** An amount included in the budget for construction, uncommitted for any specific purpose to cover the cost of unforeseen factors related to construction but not specifically addressed in the budget. Contingencies are likely for renovation, alteration, and rehabilitation projects.

**contract** An agreement between two or more individuals whereby mutual assent occurs, giving rise to a specified promise or series of promises to be performed, for which consideration is given.

**Contract documents** A term applied to any combination of related documents that collectively define the extent of an agreement between two or more parties. With regard to the contract for construction, the contract documents generally consist of the agreement (contract), bonds, certificates, conditions of the contract, specifications, drawings, and modifications.

**contract sum** An amount representing the total consideration in money to be paid the contractor for services performed under the contract for construction.

**contractor** A constructor who is a party to the contract for construction, pledged to the owner to perform the work of construction in accordance with the contract documents.

**cost plus fee (CPF)** A project-specific delivery method that compensates the contractor for total project expenditures accrued for the work, plus a stipulated fee (or fee based on a percentage of the total project cost). The CPF contractor has similar responsibilities as with a stipulated sum contract, only without the associated financial risks.

**cost plus fee with guaranteed maximum price (CPF-GMP)** A delivery method similar to CPF, except that it has a cost-capping feature that reduces the owner's risk. CPF-GMP is more popular than CPF for projects of higher value.

**critical path** A term used to describe the order of events (each of a particular duration) that results in the least amount of time required to complete a project.

**Davis-Bacon Act** An act by Congress, enacted into law in 1931, that mandates wages and fringe benefits paid to workers employed by contractors under contract with the federal government be no less than the prevailing rate for each particular trade for that location.

**design-bid-build (DBB)** A traditional method of construction project delivery involving the selection and award of professional design services followed by a separate process for construction services once the design documents are complete.

**design-build** A project-specific delivery method that allows an owner to contract with a single entity for both design and construction services. The method is well-suited for medium to large capital or renovation projects.

**design professional** A term used to describe either an architect or engineer, or both, duly licensed by state government for professional practice, who may be employed by an owner for the purpose of designing a building or other project.

**engineer** A design professional who by education, experience, and examination is duly licensed by one or more state governments for practice in the profession of engineering.

**estimator** One who is capable of predicting the probable cost of a building project.

**field order** A written modification to the contract for construction, made by the design professional, the construction administrator, or the construction manager. It documents a change to the contract documents in anticipation of the issuance of a formal change order signed by the owner, design professional, and contractor.

**field personnel** The contractor's representative staffing at the project site responsible for project management, quality assurance, quality control, supervision, subcontractor management, and materials management.

**general conditions of the contract** The part of the contract documents that defines the rights, responsibilities, and relationships between the parties to the contract.

**general requirements** Division 1 of the specifications when organized under the CSI Masterformat. In sequential sections, they detail the general administrative requirements of the project in careful coordination with the various conditions of the contract for construction.

**guaranteed maximum cost contract** A contract for construction wherein the contractor's compensation is stated as a combination of accountable cost plus a fee, with guarantee by the contractor that the total compensation will be limited to a specific amount. This type of contract may also have possible optional provisions for additional financial reward to the contractor for performance that causes total compensation to be less than the guaranteed maximum amount.

**historically underutilized business (HUB)** Businesses that are commonly referred to as minority- and/or women-owned businesses.

**indefinite delivery/indefinite quantity (ID/IQ)** Projects or tasks that may be indefinite as to start and completion dates, scope of work, and associated quantities.

**instructions to bidders** A document, part of the bidding requirements, usually prepared by the design professional. Instructions to bidders set forth specific instructions to candidate constructors on procedures, expectations of the owner, disclaimers of the owner, and other necessary information for the preparation of proposals for consideration for a competitive bid.

**invitation to bid** A written notice of an owner's intention to receive competitive bids for a construction project wherein a select group of candidates is invited to submit proposals.

**job order (JO)** A formal, written, project-specific authorization to accomplish work. Job orders are issued to the JOC contractor by the owner during the term of the JOC contract.

**job order authorization** Written authorization by the owner to accomplish the described work.

**job order contracting (JOC)** An indefinite delivery/ indefinite quantity project delivery method used for construction, remodeling, repair, and landscaping projects. The method can also be used for maintenance services. Pricing structures are based on competitively bid coefficients applied to pre-established unit prices. JOC contracts usually have options for annual renewal, which fosters partnering among parties to the contract.

**job order proposal** The design (if needed), along with a detailed scope of work including the project's performance times and price proposal, submitted by the JOC contractor to the owner in accordance with contract requirements. The scope of work and performance times are mutually agreed on before the JOC contractor submits a lump-sum, fixed-price, detailed price proposal based on the defined scope of work to the owner for consideration of approval to accomplish the defined scope of work.

**like-item method** A procedure used to price non-prepriced designated unit-price book (UPB) line items. It specifies a UPB line item that is basically the same in material function and installation as the non-prepriced line item.

**lump sum** An amount used in a contract or bid representing the total cost of compensation to the contractor to accomplish a described project or task.

**markup** A percentage of other sums that may be added to the total of all direct costs to determine a final price or contract sum (usually overhead and profit).

**maximum annual value** The owner's maximum designated amount of annual cumulative total value of projects to be authorized for accomplishment. Some JOC contracts do not identify a maximum annual value but offer flexibility to accommodate increasing volumes of work (also called potential annual value).

**minimum guaranteed annual value** The owner's minimum designated amount of annual cumulative total value of projects to be authorized for accomplishment.

**model codes** Professionally prepared building regulations and codes, regularly attended and revised, designed to be adopted by municipalities and appropriate political subdivisions by ordinance for use in regulating building construction for the welfare and safety of the general public.

**non-execution clause** A contract provision, sometimes required by the owner, advising contractors that they must respond to a JO proposal request or accomplish JOs authorized during the term of the contract, provided the JOs are in compliance with the contract's scope and intent.

**non-prepriced work** Tasks or items not covered by the UPB but within the contract's scope and general intent.

**open bidding** A common, colloquial term meaning the process of constructor selection known as competitive selection.

**partnering** A set of collaborative processes (as opposed to a relationship only); a commitment to achieve mutually identified objectives between two (or more) organizations, based on cooperation, open communication, and heightened effectiveness through continuous improvement.

**partnership** The joining of two or more individuals for a business purpose whereby profits and liabilities are shared.

**performance proposal** Sometimes referred to as the respondent's (contractor's) management/technical proposal as a component of an RFQ and/or RFP response. With JOC, this is typically weighted heavily as a criterion for contractor selection in conjunction with past performance and relevant experience.

**project record documents** The documents, certificates, and other information relating to the work, materials, products, assemblies, and equipment that the contractor is required to accumulate during construction and convey to the owner for his use prior to final payment and project closeout.

**purchase order (PO)** A written agreement to commit funds for services to be rendered and/or goods to be received. The document is commonly issued by federal or public-sector entities to service providers (contractors) or vendors.

**quality assurance/quality control plan** A written plan that addresses all aspects of quality control, including contact information for the contractor's quality control inspector responsible for surveillance of work, documentation of unsatisfactory work, corrective actions, and interface with the owner's inspectors.

**Request for a Job Order (JO) Proposal** The owner's notification to the contractor of an upcoming project (JO). This document identifies and briefly describes the project, the desired or required dates and times for performing the work, and contact information of the owner's designated project representative. The request may include other project-specific information. Response times to the request are in accordance with contract requirements.

**Request for Proposal (RFP)** The bid documents solicited to potential respondents and used to obtain contracting services. Selection is based on the highest-ranked respondent's past performance and relevant experience, management/technical proposal, price proposal, and sample JO proposal (if required). Once awarded, the RFP and the respondent's submitted proposal documents become the contract documents.

**scoping** A primary component process in response to a request for a JO proposal, which includes a site visit by the contractor's representative(s) and the owner's representative(s) to identify and document various site characteristics and job conditions that will be impacted by the project's design and/or owner's intent. Alternatives to performing work may be discussed during scoping in order to meet desired or required time lines and established budgets. With JOC, it is critical that scoping a project be as thorough and accurate as possible, since this process serves as the basis for developing the project's written scope of work.

**selective bidding** A process of competitive bidding for award of the contract for construction whereby the owner selects the constructors who are invited to bid to the exclusion of others as in the process of open bidding.

**shop drawings** Diagrams or other illustrations, schedules, performance charts, brochures, and other data prepared by the contractor or any subcontractor, manufacturer, supplier, or distributor to illustrate construction, materials, dimensions, installation, and other pertinent information for the incorporation of an element or item into the construction or a portion of the work.

**specifications** Documents that define the qualitative requirements for products, materials, and workmanship upon which the contract for construction is based.

**subcontractor** One under contract to a prime or general contractor for completion of a portion of the work for which the prime contractor is responsible.

**superintendent** The title usually applied to one who is the senior supervisor of the work at a construction site, acting on behalf of the contractor.

**term** The duration of the contract.

**time & materials (T&M)** A project delivery method involving a payment method that allows contractors to be compensated based on the total hourly labor rates bid applied to actual work time expended, plus materials purchased and equipment used. The method is best for small projects under certain circumstances, such as those requiring little or no design, together with the need for rapid mobilization.

**unit-price book (UPB)** A reliable, impartially developed, pre-established agreed-on source of unit-cost data, used as a primary component for JOC pricing structures.

**unit-price method** A pricing method typically used for repetitive work performed by a single trade or a group of closely related trades. Work is limited to the tasks identified by the owner for a specific project—or a series of projects with similar tasks. Projects can be divided into listings of specific descriptive task components, qualified as individual task line items of the work. Unit-price-based project delivery methods enable compensation to the contractor based on either the measurement of units of actual work-in-place, or the units of work estimated to be performed, with the estimate serving as a firm fixed price or lump sum.

# Index

**W**

wage rate, 82
   *Davis-Bacon*, 38, 64
walls, 100
warranties, 7
water utility, 101
West Point, 33, 34
windows, 100, 113

WinEstimator, WinEst, 46
wood and plastics, 113
work authorization, 51
work, volume of, 51
workers, labor-hours of, 26
work-in-place, units of, 27, 28,
   80, 96